PERGAMON INTE
of Science, Technology,

The 1000-volume original pap
industrial training an

Publisher: Rol

DATA
AND
FORMULAE

For Engineering Students

THIRD EDITION

DATA

AND

FORMULAE

For Engineering Students

THIRD EDITION

Compiled by the following staff of
Imperial College of Science and Technology

J. C. ANDERSON, D.Sc. (Eng.), D.I.C., C.Eng.
Professor of Electrical Materials

D. M. HUM, Ph.D., D.I.C.
Senior Lecturer in Mathematics

B. G. NEAL, Sc.D., F.Eng.
Formerly Professor of Civil Engineering

J. H. WHITELAW, D.Sc. (Eng.), Hon. D.Sc. (Lisbon), C.Eng.
Professor of Convective Heat Transfer

PERGAMON PRESS

OXFORD · NEW YORK · TORONTO · SYDNEY · PARIS · FRANKFURT

U.K.	Pergamon Press Ltd., Headington Hill Hall, Oxford OX3 0BW, England
U.S.A.	Pergamon Press Inc., Maxwell House, Fairview Park, Elmsford, New York 10523, U.S.A.
CANADA	Pergamon Press Canada Ltd., Suite 104, 150 Consumers Rd., Willowdale, Ontario M2J 1P9, Canada
AUSTRALIA	Pergamon Press (Aust.) Pty. Ltd., P.O. Box 544, Potts Point, N.S.W. 2011, Australia
FRANCE	Pergamon Press SARL, 24 rue des Ecoles, 75240 Paris, Cedex 05, France
FEDERAL REPUBLIC OF GERMANY	Pergamon Press GmbH, Hammerweg 6, D–6242 Kronberg-Taunus, Federal Republic of Germany

Copyright © 1983 B. G. Neal

First edition 1967
Second edition 1969
Reprinted 1969, 1976
Third edition 1983

Library of Congress Cataloging in Publication Data
Main entry under title:
Data and formulae for engineering students.
(Pergamon international library of science,
technology, engineering, and social studies)
1. Engineering—Handbooks, manuals, etc.
I. Anderson, J.C. (Joseph Chapman), 1922-
II. Imperial College of Science and Technology.
III. Series.
TA151.D35   1983   620'.00212   82-16499

British Library Cataloguing in Publication Data
Data and formulae for engineering students.—
3rd ed.
1. Engineering—Tables, calculations, etc.
I. Anderson, J.C.
620'.00212   TA151

ISBN 0-08-029982-2 (Hardcover)
ISBN 0-08-029981-4 (Flexicover)

In order to make this volume available as economically and as rapidly as possible the authors' typescripts have been reproduced in their original forms. This method unfortunately has its typographical limitations but it is hoped that they in no way distract the reader.

Printed in Great Britain by A. Wheaton & Co. Ltd., Exeter

PREFACE

Lecturers in engineering subjects are mainly concerned with ensuring that basic principles are thoroughly understood and can be applied correctly, and see little merit in students acquiring miscellaneous items of information such as the value of the radius of gyration of a solid sphere about a centroidal axis. However, for many students the distinction between mere factual information and the understanding of fundamentals becomes blurred, particularly near examination periods.

The compilers of this booklet therefore felt that it would be valuable to collect together those data and formulae which are often used, at an elementary level, in engineering courses. If made available in the examination room, and also to the student beforehand to enable him to become familiar with the contents, such a booklet should be helpful in defining the data and formulae which may be assumed as starting points in problems.

The booklet is intended to cover the needs of a first year undergraduate course. Beyond this level, the curricula of different engineering schools diverge so much that to meet all requirements would be well-nigh impossible. The material included may sometimes appear to be too advanced for the first year, but it must be borne in mind that the emphasis placed on subjects at different institutions varies considerably.

It is appreciated that other data are also required at first year level, notably tables giving the thermodynamic properties of fluids, section tables, and of course tables of mathematical functions. However, these are already available in other publications.

Most of the formulae in the booklet occur in the section on Mathematics. This may tempt the student to fall into the trap of believing that an understanding of mathematical methods is unnecessary, and can be replaced by a mere reliance on the use of formulae. Such an approach is quite inadequate for modern engineering courses.

The compilers feel that the Mathematical section is a useful aid to the intelligent student, for example by saving him the need for memorising numerical coefficients, provided that he then directs his efforts towards developing a full understanding of the topics listed.

During the next few years, there will be a gradual change over in the U.K. to the International System of units. This system, known as the SI system, is constructed from six basic units, namely the metre, kilogramme, second, ampere, degree Kelvin and candela. In electrical engineering, the SI system has in fact already replaced the CGS system which was formerly prevalent, and so electrical data are given only in SI units. In other branches of engineering, the British system of units, based on the foot, pound and second, is at present commonly used. Other data are therefore given in both systems of units, and an extensive list of conversion factors between the two systems has been included.

It is not yet clear which multiples or sub-multiples of certain derived SI units will eventually come into common use. For example, the derived unit of pressure or stress is the Newton per square metre (N/m^2), but this unit is inconveniently small. For pressure there are indications that the bar $(10^5 N/m^2)$ may be adopted, while for stress the Newton per square millimetre $(1N/mm^2 = 10^6 N/m^2)$ may be favoured. The relevant conversion factors and tables of data reflect these tentative choices.

The section on Science of Materials has been kept small because the modern approach to this subject is to concentrate in the early stages on showing how the mechanical and electrical properties of materials depend upon their atomic structures. Accordingly, the Periodic Table and the electronic structure of the elements are given. In addition, four elements have been singled out and values for a wide range of properties quoted, so as to give an idea of the orders of magnitude involved.

A pilot edition of the booklet was used at Imperial College in the 1966-67 Session. The experience gained has been most helpful in the preparation of this new version. Copies of the pilot edition were sent to a number of colleagues both within the College and at other Universities, and also to the examiners for Part I of the examinations of the Council of Engineering Institutions. Many valuable comments were received, as a result of which extensive modifications and additions have now been made. The compilers are deeply indebted to all those who contributed in this way, unfortunately too numerous to mention individually.

LONDON
June 1967.

J.C. Anderson
D.M. Hum
B.G. Neal
J.H. Whitelaw

Acknowledgement

The kindness of the Biometrika Trustees in permitting
reproduction of the abstracts from Biometrika Tables for
Statisticians, vol, 1, 3rd ed. (1966), which appear in
Sections 5.2, 5.3 and 5.4, is acknowledged with gratitude.

PREFACE TO SECOND EDITION

The production of a second edition of this booklet has
been necessitated by the unexpectedly high demand for the
first edition. The major change in the second edition is the
incorporation of Steam Tables in SI units, which have been
included for the convenience of students.

The only other changes are minor. In Sections 1.3 and 2
the tonne and kilopond have been included, and in the latter
section conversion factors for velocity have been added.
Additionally, the unit of stress N/mm^2 has been replaced by
the equivalent MN/m^2, the latter being a true SI unit, while
in section 9.1 the unit kN/mm^2 has been replaced by GN/m^2.
Finally, there are some small numerical changes in the data
given in Section 8.6 (i).

LONDON   J.C. Anderson
August 1968.   D.M. Hum
                                          B.G. Neal
                                          J.H. Whitelaw

Acknowledgement

The compilers thank the Controller of Her Majesty's
Stationery Office for his courtesy in granting permission to
reproduce the data contained in Section 8.8, Steam Tables in
SI units.

PREFACE TO THIRD EDITION

After 14 years' experience it was evident that some changes in content were desirable, to bring the booklet in line with accepted practice and terminology, especially in metric and SI Units.

The opportunity has been taken for extensive revision of the mathematics content (Sections 4 and 5) which now includes more material on finite difference formulae, numerical solution of differential equations and probability and statistics.

Under Thermodynamics (Section 8) a new section giving the standard atmosphere table has been added.

The numerical values, given in Section 7.3, for some of the physical and electrical properties of semiconductors have been up-dated.

LONDON
November 1982

J.C. Anderson
D.M. Hum
B.G. Neal
J.H. Whitelaw

CONTENTS

1. UNITS AND ABBREVIATIONS

1.1 Decimal prefixes

factor by which unit is multiplied	prefix	symbol
10^{12}	tera	T
10^9	giga	G
10^6	mega	M
10^3	kilo	k
10^2	hecto	h
10	deca	da
10^{-1}	deci	d
10^{-2}	centi	c
10^{-3}	milli	m
10^{-6}	micro	μ
10^{-9}	nano	n
10^{-12}	pico	p

1.2 SI units

(i) Basic units

quantity	unit	unit symbol
length	metre	m
mass	kilogram	kg
time	second	s
electric current	ampere	A
thermodynamic temperature	kelvin	K
luminous intensity	candela	cd
amount of substance	mole	mol

(ii) <u>Derived units</u>

quantity	unit	unit symbol
force	newton	$N = kg\ m/s^2$
work, energy heat	joule	$J = N\ m$
power	watt	$W = J/s$
pressure	pascal	$Pa = N/m^2$

electrical units

potential	volt	$V = W/A$
resistance	ohm	$\Omega = V/A$
charge	coulomb	$C = A\ s$
capacitance	farad	$F = A\ s/V$
electric field strength	-	V/m
electric flux density	-	C/m^2

magnetic units

magnetic flux	weber	$Wb = V\ s$
inductance	henry	$H = V\ s/A$
magnetic field strength	-	A/m
intensity of magnetisation	-	A/m
magnetic flux density	tesla	$T = Wb/m^2$

1.3 Imperial and other units

length	inch	in
	foot	ft
	angstrom	Å
	micron (micrometre)	µm
capacity	gallon	gal
	U.S. gallon	USgal
	litre	ℓ
mass	pound	lb
	gram	g
	tonne	t
force	poundal	pdl
	pound force	lbf
	ton force	tonf
	kilogram force (kilopond)	kgf(kp)
	dyne	dyn
temperature	degree Celsius	^{o}C
	degree Fahrenheit	^{o}F
	degree Rankine	^{o}R
work, energy, heat	British Thermal Unit	Btu
	I.T. calorie	cal_{IT}
	electron volt	eV
power	horse power	hp
time	hour	h

3

2. UNIT CONVERSION FACTORS

Length, volume

1 in	= 25.4 mm	1 in^3	$= 16.39 \text{ cm}^3$
1 ft	= 0.3048 m	1 ft^3	$= 0.02832 \text{ m}^3$

1 mile = 5280 ft = 1.609 km

$1 \ \mu\text{m} = 10^{-6} \text{ m} = 39.37 \ \mu\text{in}$

$1 \ \text{Å} = 10^{-10} \text{ m}$

$1 \text{ gal} = 0.1605 \text{ ft}^3 = 4.546 \ \ell$

$1 \text{ USgal} = 0.1337 \text{ ft}^3 = 3.785 \ \ell$

$1 \ \ell = 10^{-3} \text{ m}^3 = 1000 \text{ cm}^3$

Velocity

1 mile/h = 1.467 ft/s = 1.609 km/h

1 knot   = 1.689 ft/s = 1.853 km/h

Mass

1 lb   = 0.4536 kg

1 slug = 32.17 lb = 14.59 kg

1 ton   = 2240 lb = 1016 kg = 1.016 t

Density

$1 \text{ lb/in}^3 = 27.68 \text{ g/cm}^3$

$1 \text{ lb/ft}^3 = 16.02 \text{ kg/m}^3$

$1 \text{ slug/ft}^3 = 515.4 \text{ kg/m}^3$

Force

1 pdl   = 0.1383 N

1 lbf   = 32.17 pdl = 4.448 N

1 tonf   = 9.964 kN

1 kgf   = 2.205 lbf = 9.807 N

$1 \text{ dyn} = 10^{-5} \text{ N}$

Power

1 hp   = 550 ft lbf/s = 0.7457 kW

1 ft lbf/s = 1.356 W

Torque

1 lbf ft   = 1.356 N m

1 tonf ft = 3.037 kN m

4

Energy, work, heat

1 ft lbf = 1.356 J

1 kW h = 3.6 MJ

1 Btu   = 1.055 kJ = 252 cal_{IT} = 778.2 ft lbf

1 cal   = 4.187 J

1 hp h = 2.685 MJ

1 erg   = 10^{-7} J

Pressure, stress

1 lbf/in^2	= 0.07031 kgf/cm^2	= 6895 N/m^2
1 $tonf/in^2$	= 157.5 kgf/cm^2	= 15.44 N/mm^2
1 kgf/cm^2	= 0.09807 N/mm^2	= 0.9807 bar
1 kgf/mm^2	= 9.807 N/mm^2	= 0.9807 hbar
1 lbf/ft^2	= 47.88 N/m^2	= 47.88 Pa
1 ft H_2O	= 62.43 lbf/ft^2	= 2989 N/m^2
1 in Hg	= 70.73 lbf/ft^2	= 3386 N/m^2
1 mm Hg	= 1 torr	= 133.3 N/m^2
1 Int atm	= 1.013×10^5 N/m^2	= 14.70 lbf/in^2
1 bar	= 10^5 N/m^2	= 14.50 lbf/in^2

Temperature

$1^\circ C = 1.8^\circ F$

$T^\circ K = T^\circ C + 273.15^\circ C$

$T^\circ R = T^\circ F + 459.67^\circ F$

Dynamic viscosity

1 poise (g/cm s) = 0.1 kg/m s = 0.1 N s/m^2

1 kgf s/m^2 = 0.9807 kg/m s

1 lb/ft h   = 0.4134 g/m s

1 slug/ft s = 1 lbf s/ft^2 = 478.8 poise

                    = 47.88 kg/m s

1 lbf s/in^2 = 6895 kg/m s

Kinematic viscosity

1 ft^2/s = 0.09290 m^2/s

1 in^2/s = 6.452 cm^2/s

Thermal conductivity

1 Btu/ft h $^\circ$R = 1.731 J/m s K

1 cal/cm s K = 418.7 J/m s K

Electrical units

The conversion factors which follow are from the C.G.S. system to the SI system. (Note: in the C.G.S. system 1 e.m.u. = 3×10^{10} e.s.u. of charge.)

capacitance	1 e.s.u.	$= \frac{1}{9} \times 10^{-11}$ F
charge	1 e.m.u.	$= 10$ C
current	1 e.m.u.	$= 10$ A
electric field strength	1 e.s.u.	$= 3 \times 10^4$ V/m
electric flux density	1 e.s.u.	$= \frac{1}{12\pi} \times 10^{-5}$ C/m^2
electric polarisation	1 e.s.u.	$= \frac{1}{3} \times 10^{-5}$ C/m^2
inductance	1 e.m.u.	$= 10^{-9}$ H
intensity of magnetisation	1 e.m.u.	$= 10^3$ A/m
magnetic field strength	1 e.m.u.	$= \frac{1}{4\pi} \times 10^3$ A/m
magnetic flux	1 e.m.u.	$= 10^{-8}$ Wb
magnetic flux density	1 e.m.u.	$= 10^{-4}$ Wb/m^2
magnetic moment	1 e.m.u.	$= 10^{-3}$ A m^2
magnetomotive force	1 e.m.u.	$= \frac{10}{4\pi}$ A
mass susceptibility	1 e.m.u./g	$= 4\pi \times 10^{-3}$ kg^{-1}
potential	1 e.m.u.	$= 10^{-8}$ V
resistance	1 e.m.u.	$= 10^{-9}$ Ω

3. PHYSICAL CONSTANTS

Avogadro constant	N_A	$= 6.022 \times 10^{23}$/mol
Bohr magneton	μ_B	$= 9.27 \times 10^{-24}$ A m^2
Boltzmann constant	k	$= 1.381 \times 10^{-23}$ J/K
characteristic impedance of free space	Z_o	$= (\mu_o/\varepsilon_o)^{\frac{1}{2}} = 120\pi \ \Omega$
electron volt	eV	$= 1.602 \times 10^{-19}$ J
electronic charge	e	$= 1.602 \times 10^{-19}$ C
electronic rest mass	m_e	$= 9.110 \times 10^{-31}$ kg
electronic charge to mass ratio	e/m_e	$= 1.759 \times 10^{11}$ C/kg
energy for T = 290°K	kT	$= 4 \times 10^{-21}$ J
energy of ground state H atom (Rydberg energy)		$= 13.60$ eV
Faraday constant	F	$= 9.65 \times 10^{4}$ C/mol
permeability of free space	μ_o	$= 4\pi \times 10^{-7}$ H/m
permittivity of free space	ε_o	$= \dfrac{1}{36\pi} \times 10^{-9}$ F/m
Planck constant	h	$= 6.626 \times 10^{-34}$ J s
proton mass	m_p	$= 1.673 \times 10^{-27}$ kg
proton to electron mass ratio	m_p/m_e	$= 1836.2$
radius of first H orbit (Bohr atom)		$= 0.529 \times 10^{-10}$ m
standard gravitational acceleration	g	$= 9.807$ m/s^2
		$= 32.17$ ft/s^2
Stefan-Boltzmann constant	σ	$= 5.67 \times 10^{-8}$ J/m^2 s K^4
universal constant of gravitation	G	$= 6.67 \times 10^{-11}$ N m^2/kg^2
		$= 3.32 \times 10^{-11}$ lbf ft^2/lb^2
universal gas constant	R	$= 8.314$ J/mol K
velocity of light in vacuo	c	$= 2.9979 \times 10^{8}$ m/s
volume of 1 mol of ideal gas at N.T.P.		$= 22.42$ ℓ

4. MATHEMATICS

4.1 Vector algebra

$$\underline{a} = a_1\underline{i} + a_2\underline{j} + a_3\underline{k} \equiv (a_1,\ a_2,\ a_3)\ \text{etc.}$$

Scalar (dot) product:

$$\underline{a} \cdot \underline{b} = a_1 b_1 + a_2 b_2 + a_3 b_3$$

Vector (cross) product:

$$\underline{a} \times \underline{b} = \begin{vmatrix} \underline{i} & \underline{j} & \underline{k} \\ a_1 & a_2 & a_3 \\ b_1 & b_2 & b_3 \end{vmatrix}$$

Scalar triple product

$$[\underline{a}\ \underline{b}\ \underline{c}] = \underline{a} \cdot \underline{b} \times \underline{c} = \underline{a} \times \underline{b} \cdot \underline{c} = \begin{vmatrix} a_1 & a_2 & a_3 \\ b_1 & b_2 & b_3 \\ c_1 & c_2 & c_3 \end{vmatrix}$$

Repeated cross product:

$$\underline{a} \times (\underline{b} \times \underline{c}) = (\underline{a} \cdot \underline{c})\underline{b} - (\underline{a} \cdot \underline{b})\underline{c}$$

4.2 Series

$$(1+x)^{\alpha} = 1 + \alpha x + \frac{\alpha(\alpha-1)}{2!} x^2 + \frac{\alpha(\alpha-1)(\alpha-2)}{3!} x^3 + \dots ,$$

$$\text{for arbitrary } \alpha,\quad |x| < 1$$

$$e^x = 1 + x + \frac{1}{2!} x^2 + \quad \dots + \frac{1}{n!} x^n + \dots$$

$$\cos x = 1 - \frac{1}{2!} x^2 + \frac{1}{4!} x^4 - \dots + \frac{(-1)^n}{(2n)!} x^{2n} + \dots$$

$$\sin x = x - \frac{1}{3!} x^3 + \frac{1}{5!} x^5 - \dots + \frac{(-1)^n}{(2n+1)!} x^{2n+1} + \dots$$

$$\tan x = x + \frac{1}{3} x^3 + \frac{2}{15} x^5 + \frac{17}{315} x^7 + \dots$$

$$\ln(1+x) = x - \frac{1}{2} x^2 + \frac{1}{3} x^3 - \quad \dots + \frac{(-1)^n}{(n+1)} x^{n+1} + \dots$$

$$\text{for } -1 < x \leq 1$$

4.3 Trigonometric identities and hyperbolic functions

$$2 \sin a \cos b = \sin(a-b) + \sin(a+b)$$
$$2 \cos a \cos b = \cos(a-b) + \cos(a+b)$$
$$2 \sin a \sin b = \cos(a-b) - \cos(a+b)$$

$$\sin a + \sin b = 2 \sin\tfrac{1}{2}(a+b) \cos\tfrac{1}{2}(a-b)$$
$$\sin a - \sin b = 2 \cos\tfrac{1}{2}(a+b) \sin\tfrac{1}{2}(a-b)$$
$$\cos a + \cos b = 2 \cos\tfrac{1}{2}(a+b) \cos\tfrac{1}{2}(a-b)$$
$$\cos a - \cos b = -2 \sin\tfrac{1}{2}(a+b) \sin\tfrac{1}{2}(a-b)$$

$$\cos iz = \cosh z \qquad\qquad \sin iz = i \sinh z$$
$$\cosh iz = \cos z \qquad\qquad \sinh iz = i \sin z$$

4.4 Differential calculus

(i) Radius of curvature

For $s = s(\psi)$, $\qquad\qquad \rho = \left| \dfrac{ds}{d\psi} \right|$

For $y = y(x)$, $\qquad\qquad \rho = \left| \dfrac{(1 + y'^2)^{3/2}}{y''} \right|$

For $x = x(t)$, $y = y(t)$ $\qquad \rho = \left| \dfrac{(\dot{x}^2 + \dot{y}^2)^{3/2}}{\dot{x}\ddot{y} - \ddot{x}\dot{y}} \right|$

(ii) Leibniz's rule

$$D^n(fg) = f(D^n g) + \binom{n}{1}(Df)(D^{n-1}g) +$$
$$\cdots + \binom{n}{r}(D^r f)(D^{n-r}g) + \cdots + (D^n f)g.$$

(iii) <u>Taylor's expansion</u> of f(x) about x = a.

For $|h| < R$, say

$$f(a+h) = f(a) + hf'(a) + \frac{1}{2!} h^2 f''(a) +$$

$$\cdots + \frac{1}{n!} h^n f^{(n)}(a) + \varepsilon_n(h)$$

where $\varepsilon_n(h) = \frac{1}{(n+1)!} h^{n+1} f^{(n+1)}(a+\theta h)$, $0 < \theta < 1$.

(iv) <u>Taylor's series</u> for f(x,y) about the point (a,b).

For $\sqrt{(h^2 + k^2)} < R$, say

$$f(a+h,b+k) = f(a,b) + \left(h \frac{\partial f}{\partial x} + k \frac{\partial f}{\partial y} \right)_{a,b}$$

$$+ \frac{1}{2!} \left(h^2 \frac{\partial^2 f}{\partial x^2} + 2hk \frac{\partial^2 f}{\partial x \partial y} + k^2 \frac{\partial^2 f}{\partial y^2} \right)_{a,b} + \cdots$$

(v) <u>Partial differentiation</u>

(a) If $F = f(x,y)$, where $y = Y(x)$, then $F = F(x)$ and

$$\frac{dF}{dx} = \frac{\partial f}{\partial x} + \frac{\partial f}{\partial y} \frac{dY}{dx}.$$

(b) If $F = f(x,y)$, where $x = X(t)$, $y = Y(t)$ then

$$F = F(t) \quad \text{and} \quad \frac{dF}{dt} = \frac{\partial f}{\partial x} \frac{dX}{dt} + \frac{\partial f}{\partial y} \frac{dY}{dt}.$$

(c) If $F = f(x,y)$, where $x = X(u,v)$, $y = Y(u,v)$ then

$$F = F(u,v) \text{ and } \frac{\partial F}{\partial u} = \frac{\partial f}{\partial x} \frac{\partial X}{\partial u} + \frac{\partial f}{\partial y} \frac{\partial Y}{\partial u},$$

$$\frac{\partial F}{\partial v} = \frac{\partial f}{\partial x} \frac{\partial X}{\partial v} + \frac{\partial f}{\partial y} \frac{\partial Y}{\partial v}.$$

(vi) <u>Stationary points of f(x,y)</u>

Stationary points occur where $\dfrac{\partial f}{\partial x} = 0$ and $\dfrac{\partial f}{\partial y} = 0$ simultaneously.

Let (a,b) be a stationary point and write

$$\left(\frac{\partial^2 f}{\partial x^2}\right)_{a,b} = A, \qquad \left(\frac{\partial^2 f}{\partial x \partial y}\right)_{a,b} = H, \qquad \left(\frac{\partial^2 f}{\partial y^2}\right)_{a,b} = B.$$

If $H^2 - AB > 0$ then $f(x,y)$ has a <u>saddle-point</u> at (a,b).

If $H^2 - AB < 0$ and if $A < 0$ then $f(x,y)$ has a <u>maximum</u> at (a,b)

but if $A > 0$ then $f(x,y)$ has a <u>minimum</u> at (a,b).

(vii) <u>Differential equations</u>

(a) First order linear equation: $\dfrac{dy}{dx} + P(x)y = Q(x)$ has an 'integrating factor' $I(x) = e^{\int P(x)\,dx}$ so

$$\frac{d}{dx}\,(yI) = QI.$$

(b) $P(x,y)\,dx + Q(x,y)\,dy = 0$ is an 'exact equation' if

$$\frac{\partial Q}{\partial x} - \frac{\partial P}{\partial y} = 0.$$

(c) D-operator theorems: if $L(D) = a_0 + a_1 D + \cdots + a_n D^n$

$L(D)e^{kx} = L(k)e^{kx}$, where k is a constant

$L(D)[e^{kx}V(x)] = e^{kx}L(D+k)[V(x)]$, (the "shift rule")

$F(D^2){\sin \atop \cos}kx = F(-k^2){\sin \atop \cos}kx.$

11

4.5　Integral calculus

(i) <u>An important substitution</u>,　$\tan \dfrac{\theta}{2} = t$

$$\sin \theta = \frac{2t}{1+t^2} \, , \quad \cos \theta = \frac{1-t^2}{1+t^2} \, , \quad d\theta = \frac{2dt}{1+t^2} \, .$$

(ii) <u>Some indefinite integrals</u>

$$\int \sec \theta \, d\theta \; = \ln(\sec \theta + \tan \theta) = \ln \tan \left(\frac{\theta}{2} + \frac{\pi}{4} \right)$$

$$\int \csc \theta \, d\theta = \ln(\csc \theta - \cot \theta) = \ln \tan \frac{\theta}{2}$$

$$\int \frac{dx}{\sqrt{(a^2 - x^2)}} \; = \sin^{-1} \frac{x}{a} \quad \text{or,} \quad -\cos^{-1} \frac{x}{a} \, , \qquad |x| < a$$

$$\int \frac{dx}{\sqrt{(a^2 + x^2)}} \; = \sinh^{-1} \frac{x}{a} = \ln\{x + \sqrt{(a^2 + x^2)}\} - \ln a$$

$$\int \frac{dx}{\sqrt{(x^2 - a^2)}} \; = \cosh^{-1} \frac{x}{a} = \ln\{x + \sqrt{(x^2 - a^2)}\} - \ln a \, , \quad x \geq a$$

$$\int \frac{dx}{a^2 + x^2} \; = \frac{1}{a} \tan^{-1} \frac{x}{a}$$

$$\int \frac{dx}{a^2 - x^2} \; = \frac{1}{a} \tanh^{-1} \frac{x}{a} = \frac{1}{2a} \ln\left(\frac{a+x}{a-x}\right) \, , \qquad |x| < a$$

$$\int \frac{dx}{x^2 - a^2} \; = \frac{1}{2a} \ln\left(\frac{x-a}{x+a}\right) \qquad\qquad |x| > a$$

$$\int x^m \ln x \, dx = \frac{x^{m+1}}{m+1} \left[\ln x - \frac{1}{m+1} \right] \, , \quad m \neq -1$$

(iii) <u>Some definite integrals</u>

$$\int_0^{\frac{\pi}{2}} \sin^n x\,dx \; = \int_0^{\frac{\pi}{2}} \cos^n x\,dx \; = \begin{cases} \dfrac{n-1}{n} \dfrac{n-3}{n-2} \cdots \dfrac{3}{4} \dfrac{1}{2} \dfrac{\pi}{2} \, , & \text{n even} \\[2mm] \dfrac{n-1}{n} \dfrac{n-3}{n-2} \cdots \dfrac{4}{5} \dfrac{2}{3} 1 \, , & \text{n odd} \end{cases}$$

12

$$I_{m,n} = \int_0^{\frac{\pi}{2}} \sin^m x \, \cos^n x dx = \left(\frac{m-1}{m+n}\right) I_{m-2,n} = \left(\frac{n-1}{m+n}\right) I_{m,n-2} \; ,$$

$$m > 1, \quad n > 1$$

$$\int_0^{\infty} e^{-ax} \sin bx dx = \frac{b}{a^2+b^2} \; , \quad \int_0^{\infty} e^{-ax} \cos bx dx = \frac{a}{a^2+b^2} \; ,$$

$$\text{for } a > 0$$

$$\int_0^{\infty} e^{-x^2} dx = \frac{\sqrt{\pi}}{2} \; .$$

$$⨍_0^{\pi} \frac{\cos n\phi}{\cos \phi - \cos \theta} \, d\phi = \pi \frac{\sin n\theta}{\sin \theta} \; , \quad n = 0, 1, 2, \ldots$$

$$⨍_0^{\pi} \frac{\sin n\phi \, \sin \phi}{\cos \phi - \cos \theta} \, d\phi = - \pi \cos n\theta \; , \quad n = 1, 2, \ldots$$

(iv) Stirling's formula for n!

For n large, $n! \sim \sqrt{(2\pi)} \, n^{n+\frac{1}{2}} \, e^{-n}$

or, $\log_{10} n! \approx 0.39909 + (n+\frac{1}{2})\log_{10} n - 0.43429 \, n.$

4.6 Laplace Transforms

Function	Laplace Transform
$f(t)$	$F(s) = \int_0^{\infty} e^{-st} f(t) \, dt$
$af(t) + bg(t)$	$aF(s) + bG(s)$
$\dfrac{df}{dt}$	$sF(s) - f(o)$
$\dfrac{d^2f}{dt^2}$	$s^2F(s) - sf(o) - f'(o)$
$e^{at} f(t)$	$F(s-a)$

13

Function	Laplace Transform
$t^n f(t)$	$(-1)^n \dfrac{d^n F(s)}{ds^n}$ $(n=1,2,\ldots)$
$\dfrac{\partial}{\partial \alpha} f(t,\alpha)$	$\dfrac{\partial}{\partial \alpha} F(s,\alpha) = \dfrac{\partial}{\partial \alpha} \displaystyle\int_o^\infty e^{-st} f(t,\alpha)\,dt$
$\displaystyle\int_o^t f(t)\,dt$	$\dfrac{1}{s} F(s)$
$\displaystyle\int_o^t f(u)\,g(t-u)\,du$	$F(s)\ G(s)$
1	$1/s$, $s > 0$
t^n $(n=1,2,\ldots)$	$n!/s^{n+1}$, $s > 0$
e^{at}	$1/(s-a)$, $s > a$
$\sin \omega t$	$\omega/(s^2+\omega^2)$, $s > 0$
$\cos \omega t$	$s/(s^2+\omega^2)$, $s > 0$
$H(t-T) = \begin{cases} 0, & t < T \\ 1, & t > T \end{cases}$	$\dfrac{1}{s} e^{-sT}$, $T > 0$, $s > 0$
$\delta(t-a)$, $a > 0$	e^{-as}
$J_o(at)$	$(a^2+s^2)^{-\frac{1}{2}}$

4.7 Numerical analysis

(i) Approximate solution of an algebraic equation

(a) An iterative method for $x = \varphi(x)$ when $|\varphi'(x)| < 1$ near the root. If a root occurs near to $x = a$ take $x_o = a$ and

$$x_{n+1} = \varphi(x_n) \qquad \text{for } n = 0, 1, 2, \ldots$$

14

(b) If a root of $f(x) = 0$ occurs near to $x = a$, take $x_o = a$ and

$$x_{n+1} = x_n - \frac{f(x_n)}{g(x_n)} \qquad n = 0, 1, 2, \ldots$$

where <u>either</u> $g(x_n) = f'(x_n)$ (Newton tangent method)

<u>or</u> $g(x_n) = \dfrac{f(x_n) - f(x_{n-1})}{x_n - x_{n-1}}$ (Newton chord method).

(ii) <u>Least-squares fitting of a straight line</u>

If y_i ($i = 1, 2, \ldots n$) are the experimentally observed values of y at chosen (exact) values of x_i of the variable x, the line of "best fit" passes through the centroid

$$\bar{x} = \frac{1}{n} \sum_{i=1}^{n} x_i \qquad \bar{y} = \frac{1}{n} \sum_{i=1}^{n} y_i$$

and is given by $y = mx + c$ where

$$m = \frac{\Sigma(x_i - \bar{x})(y_i - \bar{y})}{\Sigma(x_i - \bar{x})^2} \quad , \quad c = \bar{y} - m\bar{x}$$

$$= \frac{\Sigma x_i y_i - n\bar{x}\bar{y}}{\Sigma x_i^2 - n\bar{x}^2} .$$

(iii) <u>Finite-difference formulae</u>

Shift operator E , $\qquad\qquad Ef(x) = f(x+h)$

Forward-difference operator Δ , $\quad \Delta f(x) = f(x+h) \quad - f(x)$

Central difference operator δ , $\quad \delta f(x) = f(x+\tfrac{1}{2}h) - f(x-\tfrac{1}{2}h)$

$$\Delta = E - 1 \qquad \delta = E^{\frac{1}{2}} - E^{-\frac{1}{2}}$$

$$E = 1 + \Delta \qquad \delta^2 = E + E^{-1} - 2$$

Symbolic form of Taylor's theorem: $E = e^{hD}$, $D = \dfrac{d}{dx}$,

i.e. $\qquad Ef(x) = (1 + hD + \dfrac{1}{2!}(hD)^2 + \ldots)f(x)$.

Forward-difference interpolation formula:

$$E^\theta f(x_o) \equiv f(x_o + \theta h) = f_o + \theta \Delta f_o + \frac{\theta(\theta-1)}{2!} \Delta^2 f_o + \dots$$

Forward-difference differentiation formula: $hD = \ln(1+\Delta)$

$$hf_o' = \Delta f_o - \frac{1}{2} \Delta^2 f_o + \frac{1}{3} \Delta^3 f_o - \dots$$

Central difference formulae for derivatives of $f(x)$.

$$f'(x) = \frac{1}{2h} [f(x+h) - f(x-h)] + \varepsilon$$
$$\text{where } \varepsilon \sim -\frac{1}{6} h^2 f'''(x) .$$

$$f''(x) = \frac{1}{2h^2} [f(x-h) + f(x+h) - 2f(x)] + \varepsilon$$
$$\text{where } \varepsilon \sim -\frac{1}{24} h^2 f^{(4)}(x) .$$

Finite difference approximation for the Laplacian of $f(x,y)$:

$$\nabla^2 f = \frac{\partial^2 f}{\partial x^2} + \frac{\partial^2 f}{\partial y^2}.$$

Writing $f_{i,j} = f(x_o + i\delta x, y_o + j\delta y)$

$$(\nabla^2 f)_{i,j} \approx \frac{1}{\delta x^2} [f_{i-1,j} + f_{i+1,j} - 2f_{i,j}]$$
$$+ \frac{1}{\delta y^2} [f_{i,j-1} + f_{i,j+1} - 2f_{i,j}].$$

(iv) <u>Lagrange's interpolation formula for unequal intervals</u>

The polynomial $P(x)$ of degree 2 passing through the three points (x_i, y_i), $i = 1, 2, 3$ is

$$P(x) = \frac{(x-x_2)(x-x_3)}{(x_1-x_2)(x_1-x_3)} y_1 + \frac{(x-x_1)(x-x_3)}{(x_2-x_1)(x_2-x_3)} y_2$$
$$+ \frac{(x-x_1)(x-x_2)}{(x_3-x_1)(x_3-x_2)} y_3$$

(v) Formulae for numerical integration

Write $\quad x_n = x_o + nh, \qquad\qquad y_n = y(x_n)$

(a) Trapezoidal Rule (1-strip):

$$\int_{x_o}^{x_1} y(x)\,dx = \frac{h}{2}\,[y_o + y_1] \;+\; \varepsilon,$$

$$\varepsilon \sim -\frac{h^3}{12}\,y_o'' \qquad \text{or,} \qquad -\frac{h}{12}\,\Delta^2\,y_o.$$

(b) Simpson's Rule (2-strip):

$$\int_{x_o}^{x_2} y(x)\,dx = \frac{h}{3}\,[y_o + 4y_1 + y_2] \;+\; \varepsilon,$$

$$\varepsilon \sim -\frac{h^5}{90}\,y_1^{iv} \qquad \text{or,} \qquad -\frac{h}{90}\,\Delta^4\,y_o.$$

(c) Milne's Rule (4-strip):

$$\int_{x_o}^{x_4} y(x)\,dx = \frac{4h}{3}\,[2y_1 - y_2 + 2y_3] \;+\; \varepsilon,$$

$$\varepsilon \sim \frac{14}{45}\,h^5 y_2^{iv} \qquad \text{or,} \qquad \frac{14}{45}\,h\Delta^4\,y_o.$$

For h small, the dominant term in the error term of each formula is given. The formulae give good approximations to the corresponding integrals provided the derivatives or differences occurring in these terms are so small as to make the error term small.

(vi) Richardson's error estimation formula for use with Simpson's rule

Let $I = \int_a^b f(x)\,dx$ and let I_1, I_2 be two estimates of I obtained by using Simpson's rule with intervals h_1 and h_2, where $h_2 < h_1$,

i.e. $\qquad\qquad h_1 = \dfrac{b-a}{n_1}, \qquad h_2 = \dfrac{b-a}{n_2} \qquad n_1,\, n_2$ even integers.

A better estimate of I is then given by

$$I = I_2 + (I_2 - I_1)/[(h_1/h_2)^4 - 1]$$

If $\quad h_2 = \tfrac{1}{2}h_1$, $I = I_2 + (I_2 - I_1)/15$.

(vii) Numerical solution of the differential equation
$\dfrac{dy}{dx} = f(x,y)$, given $y = y_o$ at $x = x_o$

$$y(x) = y_o + \int_{x_o}^{x} f(s, y(s)) \, ds \qquad [\text{Exact}]$$

(a) Second-order Runge-Kutta Formulae

Writing $x_n = x_o + nh$, $y_n = y(x_n)$

$k_1 = hf(x_n, y_n)$, $k_2 = hf(x_n + h, y_n + k_1)$,

$y_{n+1} = y_n + \tfrac{1}{2}(k_1 + k_2) + O(h^3)$. $\quad$ (n = 0, 1, ...)

(b) Fourth-order Runge-Kutta Formulae

Writing $x_n = x_o + nh$, $y_n = y(x_n)$

$k_1 = hf(x_n, y_n)$, $k_2 = hf(x_n + \tfrac{1}{2}h, y_n + \tfrac{1}{2}k_1)$,

$k_3 = hf(x_n + \tfrac{1}{2}h, y_n + \tfrac{1}{2}k_2)$, $k_4 = hf(x_n + h, y_n + k_3)$,

$y_{n+1} = y_n + \dfrac{1}{6}(k_1 + 2k_2 + 2k_3 + k_4) + \in$ (n = 0, 1, ...)

where $\varepsilon = O(h^5)$.

(c) Adams-Moulton (predictor-corrector) Formulae

Writing $x_n = x_o + nh$, $y_n = y(x_n)$, $f_n = f(x_n, y_n)$

Predictor: $y_{n+1}^{*} = y_n + \dfrac{1}{12} h[23f_n - 16f_{n-1} + 5f_{n-2}] + \varepsilon_1$

$\qquad\qquad f_{n+1}^{*} = f(x_{n+1}, y_{n+1}^{*})$

Corrector: $y_{n+1} = y_n + \dfrac{1}{12} h[5f_{n+1} + 8f_n - f_{n-1}] + \varepsilon_2$

The truncation errors terms ε_1 and ε_2 are both $O(h^4)$.

18

4.8 Fourier series

(i) General formulae

If $f(x)$ is periodic of period $2L$, $f(x+2L) = f(x)$

$$f(x) = \tfrac{1}{2}a_0 + \sum_{n=1}^{\infty} a_n \cos \frac{n\pi x}{L} + \sum_{n=1}^{\infty} b_n \sin \frac{n\pi x}{L}$$

where

$$a_n = \frac{1}{L} \int_{-L}^{L} f(x) \cos \frac{n\pi x}{L} \, dx \qquad n = 0, 1, 2, \ldots$$

$$b_n = \frac{1}{L} \int_{-L}^{L} f(x) \sin \frac{n\pi x}{L} \, dx \qquad n = 1, 2, 3, \ldots$$

If $f(x)$ is an __even__ function of x, i.e., $f(-x) = f(x)$

then

$$a_n = \frac{2}{L} \int_{0}^{L} f(x) \cos \frac{n\pi x}{L} \, dx \qquad n = 0, 1, 2, \ldots$$

and

$$b_n = 0 \qquad n = 1, 2, 3, \ldots$$

If $f(x)$ is an __odd__ function of x, i.e., $f(-x) = -f(x)$

then

$$a_n = 0 \qquad n = 0, 1, 2, \ldots$$

and

$$b_n = \frac{2}{L} \int_{0}^{L} f(x) \sin \frac{n\pi x}{L} \, dx \qquad n = 1, 2, 3, \ldots$$

(ii) Special waveforms, all of period $2L$

(a) Square wave, sine series

$$f(x) = \begin{cases} -k \quad, & -L < x < 0 \\[2ex] +k \quad, & 0 < x < L \end{cases}$$

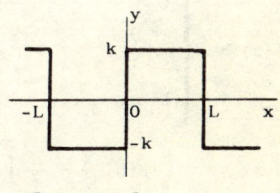

$$f(x) = \frac{4k}{\pi} \left[\sin \frac{\pi x}{L} + \frac{1}{3} \sin \frac{3\pi x}{L} + \frac{1}{5} \sin \frac{5\pi x}{L} + \ldots \right].$$

(b) <u>Square wave</u>, <u>cosine series</u>

$$f(x) = \begin{cases} k & , \quad |x| < \tfrac{1}{2}L \\ \\ -k & , \quad \tfrac{1}{2}L < |x| < L \end{cases}$$

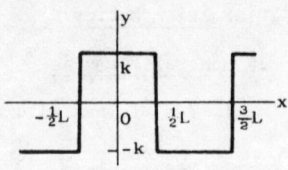

$$f(x) = \frac{4k}{\pi} \left[\cos \frac{\pi x}{L} - \frac{1}{3} \cos \frac{3\pi x}{L} + \frac{1}{5} \cos \frac{5\pi x}{L} - \dots \right].$$

(c) <u>Triangular wave</u>

$$f(x) = \begin{cases} \dfrac{2k}{L}\left(\dfrac{L}{2} + x\right) & , \quad -L < x < 0 \\ \\ \dfrac{2k}{L}\left(\dfrac{L}{2} - x\right) & , \quad 0 < x < L \end{cases}$$

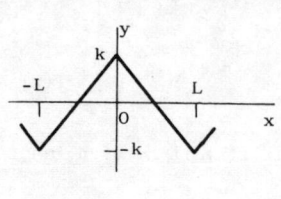

$$f(x) = \frac{8k}{\pi^2} \left[\cos \frac{\pi x}{L} + \frac{1}{3^2} \cos \frac{3\pi x}{L} + \frac{1}{5^2} \cos \frac{5\pi x}{L} + \dots \right].$$

(d) <u>Saw-tooth wave</u>

$$f(x) = \frac{kx}{L} , \quad -L < x < L$$

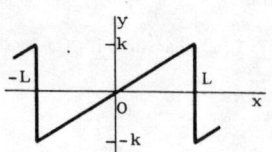

$$f(x) = \frac{2k}{\pi} \left[\sin \frac{\pi x}{L} - \frac{1}{2} \sin \frac{2\pi x}{L} + \frac{1}{3} \sin \frac{3\pi x}{L} - \dots \right].$$

(e) <u>Half-wave rectification</u>

$$f(x) = \begin{cases} 0 & , \quad -L < x < 0 \\ \\ E \sin \dfrac{\pi x}{L} & , \quad 0 < x < L \end{cases}$$

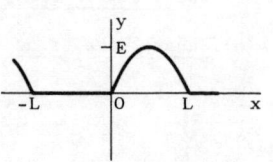

$$f(x) = \frac{E}{2} \sin \frac{\pi x}{L} + \frac{2E}{\pi} \left[\frac{1}{2} - \frac{1}{1.3} \cos \frac{2\pi x}{L} - \frac{1}{3.5} \cos \frac{4\pi x}{L} - \dots \right].$$

(f) <u>Full-wave rectification</u>

$$f(x) = \begin{cases} -E \sin \dfrac{\pi x}{L} \ , & -L < x < 0 \\[2ex] +E \sin \dfrac{\pi x}{L} \ , & 0 < x < L \end{cases}$$

$$f(x) = \frac{4E}{\pi} \left[\frac{1}{2} - \frac{1}{1.3} \cos \frac{2\pi x}{L} - \frac{1}{3.5} \cos \frac{4\pi x}{L} - \ldots \right].$$

5. PROBABILITY AND STATISTICS

5.1 Probability distributions for discrete random variables

Notation: Let R be a discrete random variable with outcome r.

Then, the probability mass function p_R is defined by

$$p_R(r) = P[R = r],$$

the expected value of R by $E[R] = \sum_{\text{all } r_i} r_i p_R(r_i)$,

and the variance of R by $V[R] = \sum_{\text{all } r_i} (r - E[R])^2 p_R(r_i)$

so that $V[R] = \sum_{\text{all } r_i} r_i^2 p_R(r_i) - (E[R])^2$.

(a) Binomial: n = number of independent trials with constant probability p of success in each,

r = number of successes.

$$p_R(r) = \binom{n}{r} p^r (1-p)^{n-r} \qquad r = 0, 1, 2, \ldots n$$

$$E[R] = np \quad , \qquad V[R] = np(1-p) .$$

(b) Poisson: μ = mean rate of occurrence of an event,

r = number of events actually occurring in unit time

$$p_R(r) = e^{-\mu} \mu^r / r! \qquad r = 0, 1, \ldots$$

$$E[R] = \mu \quad , \qquad V[R] = \mu .$$

(c) Geometric: p = constant probability of success in a sequence of trials,

r = number of trials up to and including first success

$$p_R(r) = (1-p)^{r-1} p \qquad r = 1, 2, \ldots$$

$$E[R] = 1/p \quad , \qquad V[R] = (1-p)/p^2 .$$

22

5.2 Probability distributions for continuous random variables

Notation: Let X be a continuous random variable with outcome x.

Then, the probability distribution function F_X is defined by

$$F_X(x) = P[X \leq x]$$

and the probability density function f_X by

$$P[x_1 \leq X \leq x_2] = \int_{x_1}^{x_2} f_X(x)dx ,$$

so that $f_X(x) = \dfrac{d}{dx} F_X(x)$,

the expected value of X by $E[X] = \displaystyle\int_{-\infty}^{+\infty} x f_X(x)dx$

and the variance of X by $V[X] = \displaystyle\int_{-\infty}^{+\infty} (x-E[X])^2 f_X(x)dx$,

so that $V[X] = \displaystyle\int_{-\infty}^{+\infty} x^2 f_X(x)dx - (E[X])^2$.

(a) Exponential:

$$f_X(x) = \lambda e^{-\lambda x} , \qquad\qquad x \geq 0 , \quad \lambda > 0 ,$$

$$E[X] = 1/\lambda , \qquad\qquad V[X] = 1/\lambda^2 .$$

(b) Normal: the standardised normal distribution, $N(0,1)$

has $\qquad\qquad f_Z(z) = \dfrac{1}{\sqrt{2\pi}} e^{-z^2/2}$

$$E[Z] = 0 , \qquad\qquad V[Z] = 1 .$$

$\Phi(z') =$ probability that the random variable Z is observed to have a value $\leq z'$ (the shaded area shown) is given by the following table:

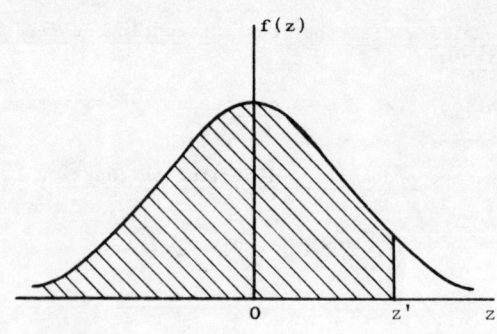

z'	$\Phi(z')$	z'	$\Phi(z')$	z'	$\Phi(z')$
0.0	.5000	1.0	.8413	2.0	.9772
.1	.5398	.1	.8643	.1	.9821
.2	.5793	.2	.8849	.2	.9861
.3	.6179	.3	.9032	.3	.9893
.4	.6554	.4	.9192	.4	.9918
0.5	.6915	1.5	.9332	2.5	.9938
.6	.7257	.6	.9452	.6	.9953
.7	.7580	.7	.9554	.7	.9965
.8	.7881	.8	.9641	.8	.9974
.9	.8159	.9	.9713	.9	.9981

Percentage points of the Normal Distribution N(0,1)

$\Phi(z')$	%(1-tail)	%(2-tail)	z'
.9500	5.0	10	1.6449
.9750	2.5	5	1.9600
.9900	1.0	2	2.3263
.9950	0.5	1	2.5758

The <u>general normal distribution</u> $N(\mu, \sigma^2)$ has

$$f_X(x) = \frac{1}{\sigma\sqrt{2\pi}} e^{-(x-\mu)^2/2\sigma^2} \ ,$$

$$E[X] = \mu \qquad , \qquad V[X] = \sigma^2 \ .$$

To use tables of $\Phi(z)$, take $z = \frac{x-\mu}{\sigma}$.

eg $\{a < x < b\} = \Phi_b - \Phi_a$

(c) <u>Gamma</u>

$$f_X(x) = \frac{\beta^\alpha x^{\alpha-1} e^{-\beta x}}{\Gamma(\alpha)} \ , \qquad 0 < x < \infty \qquad (\alpha > 0 \ , \ \beta > 0)$$

$$E[X] = \frac{\alpha}{\beta} \qquad , \qquad V[X] = \frac{\alpha}{\beta^2} \ .$$

5.3 Confidence interval for mean of a normal distribution

Let x_1, x_2, ... x_n denote a set of n observations of a
random variable X having a normal distribution whose
mean μ is unknown. Then,

Sample mean $\bar{x} = \frac{1}{n}\sum x_i$

Sample standard deviation $= s$, $s^2 = \frac{1}{n-1}\sum(x_i - \bar{x})^2$

Distribution of x is $N(\mu, \sigma^2)$

Distribution of $\bar{x}$ is $N(\mu, \sigma^2/n)$

Distribution of $\dfrac{\bar{x} - \mu}{(\sigma/\sqrt{n})}$ is $N(0,1)$.

<u>If the variance σ^2 is known</u>, the 95% confidence interval
for μ is

$$\left(\bar{x} - 1.96 \frac{\sigma}{\sqrt{n}} \quad , \quad \bar{x} + 1.96 \frac{\sigma}{\sqrt{n}} \right)$$

which is obtained from

$$\Pr\left\{ -1.96 < \frac{\bar{x} - \mu}{\sigma/\sqrt{n}} < 1.96 \right\} = 0.95 \ .$$

If the variance σ^2 is unknown, $\dfrac{\bar{x}-\mu}{s/\sqrt{n}}$ has the
t-distribution with n-1 degrees of freedom. The 95%
confidence interval for μ is

$$\left(\bar{x} - t_c \frac{s}{\sqrt{n}} \quad , \quad \bar{x} + t_c \frac{s}{\sqrt{n}} \right)$$

which is obtained from

$$\Pr \left\{ - t_c < \frac{\bar{x}-\mu}{s/\sqrt{n}} < t_c \right\} = 0.95 .$$

Values of t_c for various degrees of freedom are given
below

95% points of the t-distribution

n-1	t_c	n-1	t_c	n-1	t_c
1	12.7	6	2.45	12	2.18
2	4.30	7	2.36	15	2.13
3	3.18	8	2.31	20	2.09
4	2.78	9	2.26	30	2.04
5	2.57	10	2.23	60	2.00
				∞	1.96

5.4 Goodness of fit - the χ^2 test

Let k be the number of mutually exclusive possible
outcomes. If O_1, O_2, ... O_k are the observed frequencies
of the outcomes and, under a specified hypothesis, the
expected frequencies are E_1, E_2, ... E_k, then

$$S = \sum_{i=1}^{k} \frac{(O_i - E_i)^2}{E_i}$$

is approximately χ^2-distributed with k-m degrees of
freedom, where m is the number of known relations among
the expected frequencies. Since $P[S \geq \chi^2_{\alpha,\nu}] = \alpha$, the
hypothesis is rejected at significance level α if
$S \geq \chi^2_{\alpha,\nu}$.

Critical values of X^2 for 5% and 1% points

ν	5%	1%	ν	5%	1%	ν	5%	1%
1	3.84	6.63	11	19.68	24.73	22	33.92	40.29
2	5.99	9.21	12	21.03	26.22	24	36.42	42.98
3	7.81	11.34	13	22.36	27.69	26	38.89	45.64
4	9.49	13.28	14	23.68	29.14	28	41.34	48.28
5	11.07	15.09	15	25.00	30.58	30	43.77	50.89
6	12.59	16.81	16	26.30	32.00	40	55.76	63.69
7	14.07	18.48	17	27.59	33.41	50	67.50	76.15
8	15.51	20.09	18	28.87	34.81	60	79.08	88.38
9	16.92	21.67	19	30.14	36.19	80	101.9	112.3
10	18.31	23.21	20	31.41	37.57	100	124.3	135.8

5.5 Linear regression

$$E[Y_i] = a(x_i - \bar{x}) + b \quad , \quad V[Y_i] = \sigma^2$$

where values x_i are known and a, b are unknown.

$$\hat{a} = \frac{\Sigma(x_i - \bar{x})(Y_i - \bar{Y})}{\Sigma(x_i - \bar{x})^2} \quad , \quad \hat{b} = \bar{Y}$$

$$V[\hat{a}] = \frac{\sigma^2}{\Sigma(x_i - \bar{x})^2} \quad , \quad V[\hat{b}] = \frac{\sigma^2}{n}.$$

5.6 Poisson process, rate μ

A Poisson process is a random sequence of events satisfying either of the following equivalent conditions:

(i) the times between the events are independent and exponentially distributed with parameter μ,

(ii) the number of events in a time interval of length t has a Poisson distribution with mean μt.

27

6.   PROPERTIES OF LAMINAE AND SOLIDS

6.1   Pappus's theorems

A solid of revolution is formed by rotating a closed area about an axis external to the area.

(a) The volume of the solid is equal to the area multiplied by the length of the path traced out by the centroid of the area.

(b) The surface area of the solid is equal to the perimeter of the closed area multiplied by the length of the path traced out by the centroid of the perimeter.

6.2   Moments of inertia: general theorems

Solids or laminae: Parallel axis theorem

$$I_{xx} = I_{XX} + m\bar{y}^2$$

$$I_{xy} = I_{XY} + m\bar{x}\bar{y}$$

$$I_{yy} = I_{YY} + m\bar{x}^2$$

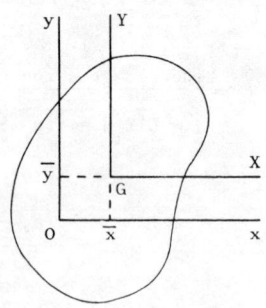

Laminae: Perpendicular axis theorem

$$I_{zz} = J_o = I_{xx} + I_{yy}$$

6.3   Radii of gyration

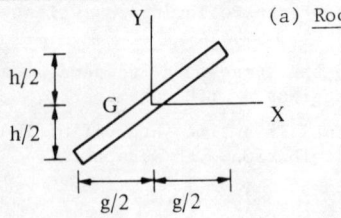

(a) Rod

$$k_{xx}^2 = \frac{1}{12}\, h^2$$

$$k_{xy}^2 = \frac{1}{12}\, gh$$

$$k_{yy}^2 = \frac{1}{12}\, g^2$$

28

(b) <u>Laminae</u>

$$k_{XX}^2 \qquad k_{YY}^2$$

Rectangle

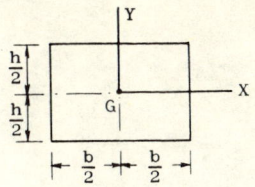

$$\frac{1}{12}\, h^2 \qquad \frac{1}{12}\, b^2$$

Circle

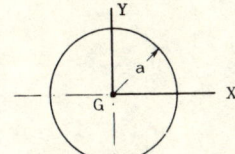

$$\frac{1}{4}\, a^2 \qquad \frac{1}{4}\, a^2$$

Semi-circle

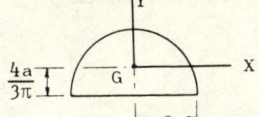

$$a^2\left[\frac{1}{4} - \left(\frac{4}{3\pi}\right)^2\right] \qquad \frac{1}{4}\, a^2$$

Triangle

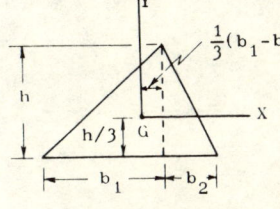

$$\frac{1}{18}\, h^2 \qquad \frac{1}{18}(b_1^2 + b_1 b_2 + b_2^2)$$

$$k_{XY}^2 = \frac{1}{36}\, h(b_1 - b_2)$$

(c) <u>Solids</u>

$$k^2_{XX} \qquad k^2_{YY}$$

Cylinder

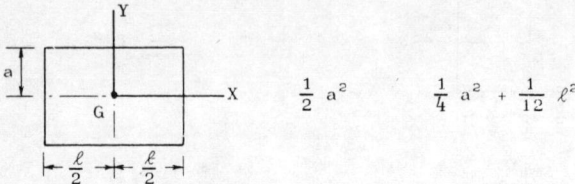

$$\frac{1}{2}\,a^2 \qquad \frac{1}{4}\,a^2 + \frac{1}{12}\,\ell^2$$

Thick-walled cylinder

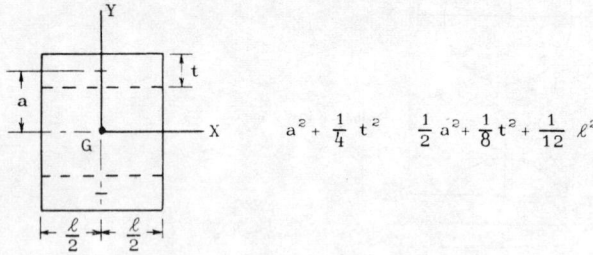

$$a^2 + \frac{1}{4}\,t^2 \qquad \frac{1}{2}\,a^2 + \frac{1}{8}\,t^2 + \frac{1}{12}\,\ell^2$$

Sphere

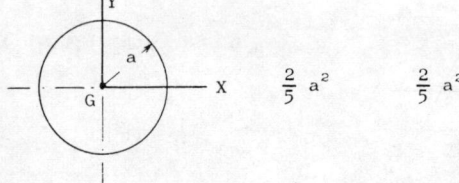

$$\frac{2}{5}\,a^2 \qquad \frac{2}{5}\,a^2$$

7. SCIENCE OF MATERIALS

7.1 Periodic Table of the Elements

I	O
1.008 H 1	4.003 He 2

I	II	III	IV	V	VI	VII	O
6.940 Li 3	9.012 Be 4	10.811 B 5	12.011 C 6	14.007 N 7	15.999 O 8	18.998 F 9	20.183 Ne 10
22.990 Na 11	24.312 Mg 12	26.982 Al 13	28.086 Si 14	30.974 P 15	32.064 S 16	35.453 Cl 17	39.948 A 18

IA	IIA	IIIA	IVA	VA	VIA	VIIA	VIII			IB	IIB	IIIB	IVB	VB	VIB	VIIB	O
39.102 K 19	40.080 Ca 20	44.956 Sc 21	47.900 Ti 22	50.942 V 23	51.996 Cr 24	54.938 Mn 25	55.847 Fe 26	58.933 Co 27	58.710 Ni 28	63.540 Cu 29	65.370 Zn 30	69.720 Ga 31	72.590 Ge 32	74.922 As 33	78.960 Se 34	79.909 Br 35	83.800 Kr 36
85.470 Rb 37	87.620 Sr 38	88.905 Y 39	91.220 Zr 40	92.906 Nb 41	95.940 Mo 42	(99) Tc 43	101.070 Ru 44	102.905 Rh 45	106.400 Pd 46	107.870 Ag 47	112.400 Cd 48	114.820 In 49	118.690 Sn 50	121.750 Sb 51	127.600 Te 52	126.904 I 53	131.300 Xe 54
132.905 Cs 55	137.340 Ba 56	*	178.490 Hf 72	180.948 Ta 73	183.850 W 74	186.200 Re 75	190.2 Os 76	192.2 Ir 77	195.09 Pt 78	196.967 Au 79	200.590 Hg 80	204.370 Tl 81	207.190 Pb 82	208.980 Bi 83	(210) Po 84	(210) At 85	(222) Rn 86
(223) Fr 87	(227) Ra 88	(227) Ac 89	232.038 Th 90	(231) Pa 91	238.030 U 92	(237) Np 93	(242) Pu 94										

* Rare earths: 57La, 58Ce, 59Pr, 60Nd, 61Pm, 62Sm, 63Eu, 64Gd, 65Tb, 66Dy, 67Ho, 68Er, 69Tm, 70Yb, 71Lu

7.2 Electronic structure of the elements

Element and Atomic Number	K	L		M			N			
	s	s	p	s	p	d	s	p	d	f
1 H	1									
2 He	2									
3 Li	2	1								
4 Be	2	2								
5 B	2	2	1							
6 C	2	2	2							
7 N	2	2	3							
8 O	2	2	4							
9 F	2	2	5							
10 Ne	2	2	6							
11 Na	2	2	6	1						
12 Mg	2	2	6	2						
13 Al	2	2	6	2	1					
14 Si	2	2	6	2	2					
15 P	2	2	6	2	3					
16 S	2	2	6	2	4					
17 Cl	2	2	6	2	5					
18 A	2	2	6	2	6					
19 K	2	2	6	2	6		1			
20 Ca	2	2	6	2	6		2			
21 Sc	2	2	6	2	6	1	2			
22 Ti	2	2	6	2	6	2	2			

Element and Atomic Number	K	L	M			N				O		
	–	–	s	p	d	s	p	d	f	s	p	d
23 V	2	8	2	6	3	2						
24 Cr	2	8	2	6	5	1						
25 Mn	2	8	2	6	5	2						
26 Fe	2	8	2	6	6	2						
27 Co	2	8	2	6	7	2						
28 Ni	2	8	2	6	8	2						
29 Cu	2	8	2	6	10	1						
30 Zn	2	8	2	6	10	2						
31 Ga	2	8	2	6	10	2	1					
32 Ge	2	8	2	6	10	2	2					
33 As	2	8	2	6	10	2	3					
34 Se	2	8	2	6	10	2	4					
35 Br	2	8	2	6	10	2	5					
36 Kr	2	8	2	6	10	2	6					
37 Rb	2	8	2	6	10	2	6			1		
38 Sr	2	8	2	6	10	2	6			2		
39 Y	2	8	2	6	10	2	6	1		2		
40 Zr	2	8	2	6	10	2	6	2		2		
41 Nb	2	8	2	6	10	2	6	4		1		
42 Mo	2	8	2	6	10	2	6	5		1		
43 Tc	2	8	2	6	10	2	6	6		1		
44 Ru	2	8	2	6	10	2	6	7		1		
45 Rh	2	8	2	6	10	2	6	8		1		
46 Pd	2	8	2	6	10	2	6	10				
47 Ag	2	8	2	6	10	2	6	10		1		
48 Cd	2	8	2	6	10	2	6	10		2		

Element and Atomic Number	K	L	M	N				O			P
	–	–	–	s	p	d	f	s	p	d	s
49 In	2	8	18	2	6	10		2	1		
50 Sn	2	8	18	2	6	10		2	2		
51 Sb	2	8	18	2	6	10		2	3		
52 Te	2	8	18	2	6	10		2	4		
53 I	2	8	18	2	6	10		2	5		
54 Xe	2	8	18	2	6	10		2	6		
55 Cs	2	8	18	2	6	10		2	6		1
56 Ba	2	8	18	2	6	10		2	6		2
57 La	2	8	18	2	6	10		2	6	1	2
58 Ce	2	8	18	2	6	10	2	2	6		2
59 Pr	2	8	18	2	6	10	3	2	6		2
60 Nd	2	8	18	2	6	10	4	2	6		2
61 Pm	2	8	18	2	6	10	5	2	6		2
62 Sm	2	8	18	2	6	10	6	2	6		2
63 Eu	2	8	18	2	6	10	7	2	6		2
64 Gd	2	8	18	2	6	10	7	2	6	1	2
65 Tb	2	8	18	2	6	10	9	2	6		2
66 Dy	2	8	18	2	6	10	10	2	6		2
67 Ho	2	8	18	2	6	10	11	2	6		2
68 Er	2	8	18	2	6	10	12	2	6		2
69 Tm	2	8	18	2	6	10	13	2	6		2
70 Yb	2	8	18	2	6	10	14	2	6		2
71 Lu	2	8	18	2	6	10	14	2	6	1	2
72 Hf	2	8	18	2	6	10	14	2	6	2	2
73 Ta	2	8	18	2	6	10	14	2	6	3	2
74 W	2	8	18	2	6	10	14	2	6	4	2

Element and Atomic Number	K	L	M	N	O				P			Q
	−	−	−	−	s	p	d	f	s	p	d	s
75 Re	2	8	18	32	2	6	5		2			
76 Os	2	8	18	32	2	6	6		2			
77 Ir	2	8	18	32	2	6	7		2			
78 Pt	2	8	18	32	2	6	8		2			
79 Au	2	8	18	32	2	6	10		1			
80 Hg	2	8	18	32	2	6	10		2			
81 Tl	2	8	18	32	2	6	10		2	1		
82 Pb	2	8	18	32	2	6	10		2	2		
83 Bi	2	8	18	32	2	6	10		2	3		
84 Po	2	8	18	32	2	6	10		2	4		
85 At	2	8	18	32	2	6	10		2	5		
86 Rn	2	8	18	32	2	6	10		2	6		
87 Fr	2	8	18	32	2	6	10		2	6		1
88 Ra	2	8	18	32	2	6	10		2	6		2
89 Ac	2	8	18	32	2	6	10		2	6	1	2

The exact electronic configuration of the later elements is uncertain

Element	Th	Pa	U	Np	Pu	Am	Cm	Bk	Cf	Es	Fm	Md	No	
Atomic Number	90	91	92	93	94	95	96	97	98	99	100	101	102	103

7.3 Some typical values of physical properties

All values are given, unless otherwise stated, for a temperature of $20^{\circ}C$.

(i) Metals

property	copper	iron
crystal structure	f.c.c.	b.c.c.
lattice constant (nm)	0.361	0.286
density (kg/m^3)	8.96×10^3	7.87×10^3
atomic volume (m^3/mol)	7.09×10^{-6}	7.10×10^{-6}
bonding	metallic	metallic
resistivity (Ω m)	1.72×10^{-8}	10×10^{-8}
cohesive energy (J/mol)	3.38×10^5	4.05×10^5
melting point ($^{\circ}C$)	1083	1530
coefficient of linear expansion (per $^{\circ}C$)	16.7×10^{-6}	12.1×10^{-6}
Fermi energy (eV)	7.04	11.2
work function (eV)	4.07–4.18	3.91–4.77
temperature coefficient of resistance (per $^{\circ}C$)	+0.0043	+0.0065
effective radius (nm) of		
(a) neutral atom	0.127	0.126
(b) singly charged ion	0.096	–
(c) doubly charged ion	0.070	0.075

(ii) Semiconductors

property	germanium	silicon
crystal structure	diamond	diamond
lattice constant (nm)	0.56575	0.54307
density (kg/m^3)	5.32×10^3	2.33×10^3
atomic volume (m^3/mol)	13.5×10^{-6}	12.0×10^{-6}
bonding	covalent	covalent
cohesive energy (J/mol)	3.72×10^5	4.39×10^5
melting point ($^{\circ}C$)	958.5	1412

Hall mobility ($m^2/V \cdot s$) (at 300 K)	electrons 0.38	electrons 0.16
	holes 0.18	holes 0.05
energy gap (eV) (room temperature)	0.67	1.14
density of states isotropic average effective mass:	electrons 0.12 m_e	electrons 0.26 m_e
	holes 0.32 m_e	holes 0.50 m_e
relative permittivity	16.0	11.8

DFE - D

8. THERMODYNAMICS AND FLUID MECHANICS

8.1 Nomenclature

quantity	symbol
specific heat at constant pressure	c_p
specific heat at constant volume	c_v
coefficient of mass diffusion	$\mathcal{D}$
acceleration due to gravity	g
specific enthalpy	h
heat transfer coefficient	h'
thermal conductivity	k
characteristic length	L
pressure	p
gas constant	R
entropy	S
time	t
temperature	T
temperature difference	ΔT
specific internal energy	u
internal energy	U
volume	V
velocity	$\mathcal{V}$
velocity of sound	a
coefficient of volumetric expansion	β
viscosity	μ
density	ρ
surface tension in contact with air	σ

8.2   Thermodynamic definitions

enthalpy	$H \equiv U + pV$
Helmholtz function	$A \equiv U - TS$
Gibbs function	$G \equiv H - TS$
specific heat at constant volume	$c_v \equiv \left(\dfrac{\partial u}{\partial T}\right)_v$
specific heat at constant pressure	$c_p \equiv \left(\dfrac{\partial h}{\partial T}\right)_p$

8.3   Dimensionless groups

Fourier number	$\equiv (k/\rho c_p)\,t/L^2$
Froude number	$\equiv \mathcal{v}^2/Lg$
Grashof number	$\equiv g\beta\Delta TL^3\rho^2/\mu^2$
Mach number	$\equiv \mathcal{v}/a$
Nusselt number	$\equiv h'L/k$
Prandtl number	$\equiv \mu c_p/k$
Reynolds number	$\equiv \rho\mathcal{v}L/\mu$
Schmidt number	$\equiv \mu/\rho\,\mathcal{D}$

8.4   Temperatures at the primary fixed points

Normal boiling point of oxygen (oxygen point)	-182.97°C
Triple point of water	0.01°C
Normal boiling point of water (steam point)	100.00°C
Normal boiling point of sulphur (sulphur point)	444.6°C
Normal melting point of silver (silver point)	960.8°C
Normal melting point of gold (gold point)	1063°C

8.5 Critical constants

	molecular weight	$T_c(K)$	p_c(bar)	ρ_c(kg/m^3)
hydrogen	2.02	33.3	13.0	31
helium (4)	4.00	5.3	2.29	69.3
water vapour	18.02	647.30	221.2	318.3
nitrogen	28.01	126.1	33.9	311
oxygen	32.00	154.4	50.4	430
carbon dioxide	44.01	304.15	73.8	468

8.6 Approximate physical properties at 20°C, 1 bar

Units

c_p	kJ/kg K	μ	kg/m s
R	kJ/kg K	σ	N/m
ρ	kg/m^3	k	kJ/m s K

(i) gases

	R	ρ	c_p	c_p/c_v	μ	k
hydrogen	4.16	0.082	14.3	1.40	8.8×10^{-6}	1.8×10^{-4}
helium	2.08	0.164	5.23	1.66	1.96×10^{-5}	1.4×10^{-4}
nitrogen	0.294	1.16	1.04	1.40	1.76×10^{-5}	2.6×10^{-5}
oxygen	0.260	1.31	0.91	1.40	2.03×10^{-5}	2.6×10^{-5}
carbon dioxide	0.190	1.80	0.84	1.28	1.47×10^{-5}	1.7×10^{-5}
air	0.287	1.19	1.00	1.40	1.82×10^{-5}	2.6×10^{-5}

40

(ii) liquids

	ρ	c_p	μ	k	σ
water	1000	4.19	1.002×10^{-3}	6.0×10^{-4}	0.073
mercury	13,600	0.14	1.55×10^{-3}	8.7×10^{-3}	0.51
castor oil	960	2.20	9.86×10^{-1}	1.8×10^{-4}	0.039
benzene	880	1.80	6.56×10^{-4}	1.6×10^{-4}	0.029
ethyl alcohol	790	2.86	1.20×10^{-3}	1.9×10^{-4}	0.022
engine oil	890	1.88	8×10^{-2}	1.5×10^{-4}	-
Freon 12	1350	0.96	2.73×10^{-4}	7.3×10^{-5}	-

(iii) solids

	ρ	c_p	k
duralumin	2720	0.88	1.7×10^{-1}
mild steel	7850	0.46	5.2×10^{-2}
stainless steel (18% Ni, 8% Cr)	7810	0.46	1.6×10^{-2}
brass	8410	0.38	1.05×10^{-1}
concrete	2400	0.88	1.1×10^{-3}
wood (pine)	500	2.8	1.5×10^{-4}
firebrick	170	0.81	3.8×10^{-4}

8.7 Friction factor

Chart for fully-developed flow of a uniform-property Newtonian fluid in straight pipes of circular cross-section, diameter d.

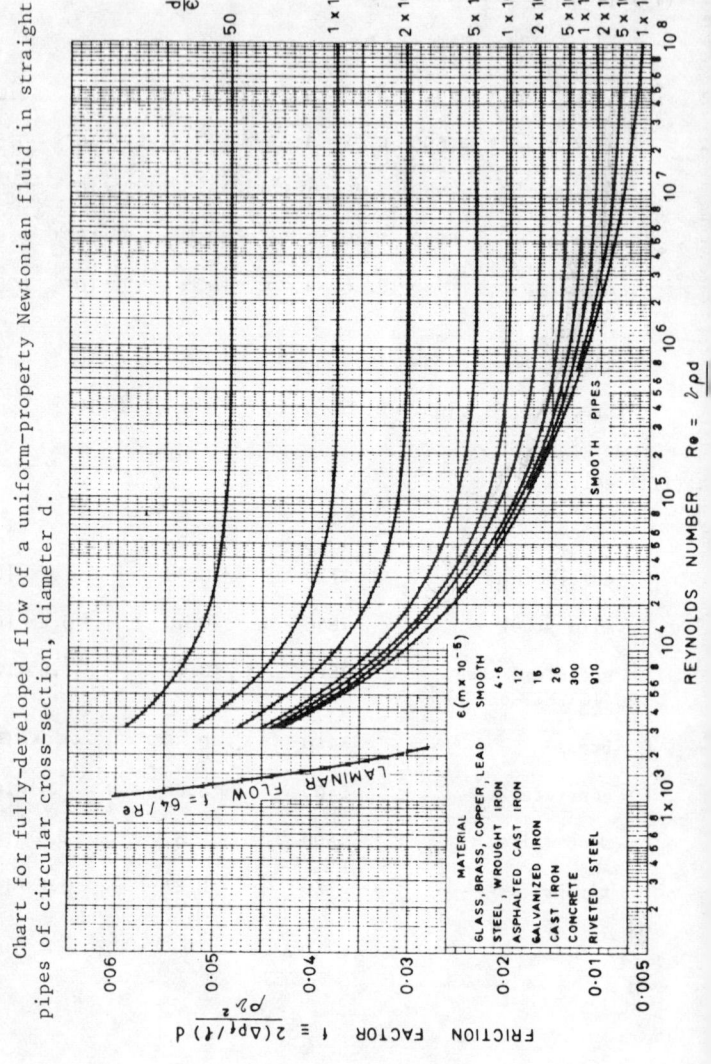

42

8.8   Steam tables in SI units

Symbols and Units

quantity	symbol	unit
temperature	T	$^{\circ}C$
absolute pressure	p	bar
specific volume	v	m^3/kg
specific volume of saturated liquid	v_f	m^3/kg
specific volume of saturated vapour	v_g	m^3/kg
specific enthalpy	h	kJ/kg
specific enthalpy of saturated liquid	h_f	kJ/kg
specific latent heat of evaporation	h_{fg}	kJ/kg
specific enthalpy of saturated vapour	h_g	kJ/kg
specific entropy	s	$kJ/kg\ K$
specific entropy of saturated liquid	s_f	$kJ/kg\ K$
specific latent entropy of evaporation	s_{fg}	$kJ/kg\ K$
specific entropy of saturated vapour	s_g	$kJ/kg\ K$

The above units comply with the SI system of units except for the units of pressure and temperature. The tables are abridged from the NEL Steam Tables 1964 which use bar and $^{\circ}C$ as the units of pressure and temperature respectively.

43

PROPERTIES OF SATURATED WATER AND STEAM : TEMPERATURE TABLE

T	p	v_f	v_g	h_f	h_{fg}	h_g	s_f	s_{fg}	s_g
0.01	0.006107	0.001000	206.3	-0.04	2500.8	2500.8	-0.0002	9.155	9.155
0.01	0.006112	0.001000	206.1	+0.00	2500.8	2500.8	+0.0000	9.155	9.155
2	0.007054	0.001000	179.9	8.4	2496.0	2504.3	0.031	9.071	9.102
4	0.008129	0.001000	157.3	16.8	2491.3	2508.1	0.061	8.989	9.050
6	0.009346	0.001000	137.8	25.2	2486.6	2511.8	0.091	8.908	8.999
8	0.010721	0.001000	121.0	33.6	2481.9	2515.5	0.121	8.828	8.949
10	0.012271	0.001000	106.4	42.0	2477.2	2519.2	0.151	8.749	8.900
12	0.014015	0.001001	93.83	50.4	2472.5	2522.9	0.181	8.671	8.851
14	0.015974	0.001001	82.89	58.8	2467.8	2526.6	0.210	8.594	8.804
16	0.018170	0.001001	73.38	67.1	2463.1	2530.2	0.239	8.518	8.757
18	0.020626	0.001001	65.08	75.5	2458.4	2533.9	0.268	8.444	8.711
20	0.023368	0.001002	57.84	83.9	2453.7	2537.5	0.296	8.370	8.666
25	0.031663	0.001003	43.40	104.8	2441.9	2546.6	0.367	8.190	8.557
30	0.042418	0.001004	32.93	125.7	2430.0	2555.7	0.437	8.016	8.452
35	0.056217	0.001006	25.25	146.6	2418.1	2564.7	0.505	7.847	8.352
40	0.073750	0.001008	19.55	167.5	2406.2	2573.7	0.572	7.684	8.256
45	0.095818	0.001010	15.28	188.4	2394.2	2582.6	0.638	7.525	8.164
50	0.12335	0.001012	12.05	209.3	2382.2	2591.4	0.704	7.372	8.075
55	0.15740	0.001015	9.578	230.2	2370.1	2600.3	0.768	7.223	7.990
60	0.19919	0.001017	7.678	251.1	2357.9	2609.0	0.831	7.078	7.909
65	0.25008	0.001020	6.201	272.0	2345.6	2617.7	0.893	6.937	7.830
70	0.31160	0.001023	5.045	293.0	2333.3	2626.3	0.955	6.800	7.754
75	0.38547	0.001026	4.133	313.9	2320.9	2634.8	1.015	6.666	7.682
80	0.47359	0.001029	3.408	334.9	2308.3	2643.2	1.075	6.536	7.611
85	0.57803	0.001032	2.828	355.9	2295.6	2651.5	1.134	6.410	7.544
90	0.70109	0.001036	2.361	376.9	2282.8	2659.7	1.193	6.286	7.478
95	0.84526	0.001040	1.982	398.0	2269.8	2667.8	1.250	6.166	7.416

T	p	v_f	v_g	h_f	h_{fg}	h_g	s_f	s_{fg}	s_g
100	1.0133	0.001044	1.673	419.1	2256.7	2675.8	1.307	6.048	7.355
105	1.2080	0.001047	1.419	440.2	2243.5	2683.6	1.363	5.933	7.296
110	1.4326	0.001052	1.210	461.3	2230.0	2691.3	1.419	5.820	7.239
115	1.6905	0.001056	1.037	482.5	2216.4	2698.9	1.473	5.710	7.183
120	1.9853	0.001060	0.8917	503.0	2202.5	2706.3	1.528	5.602	7.130
125	2.3209	0.001065	0.7704	525.0	2188.5	2713.5	1.581	5.497	7.078
130	2.7012	0.001070	0.6683	546.3	2174.2	2720.5	1.634	5.393	7.027
135	3.1305	0.001075	0.5820	567.7	2159.7	2727.3	1.687	5.291	6.978
140	3.6136	0.001080	0.5087	589.1	2144.8	2733.9	1.739	5.191	6.930
145	4.1549	0.001085	0.4461	610.6	2129.8	2740.4	1.791	5.093	6.884
150	4.7597	0.001091	0.3926	632.2	2114.4	2746.5	1.842	4.997	6.838
155	5.4331	0.001096	0.3466	653.8	2098.7	2752.5	1.892	4.902	6.794
160	6.1805	0.001102	0.3069	675.5	2082.7	2758.1	1.942	4.808	6.751
165	7.0076	0.001108	0.2725	697.3	2066.3	2763.7	1.992	4.716	6.708
170	7.9203	0.001114	0.2426	719.1	2049.6	2768.7	2.042	4.625	6.667
175	8.9247	0.001121	0.2166	741.1	2032.4	2773.5	2.091	4.535	6.626
180	10.027	0.001128	0.1939	763.1	2014.9	2778.0	2.139	4.446	6.586
185	11.234	0.001134	0.1739	785.1	1997.2	2782.3	2.188	4.359	6.546
190	12.552	0.001142	0.1564	807.5	1978.6	2786.1	2.236	4.272	6.508
195	13.988	0.001149	0.1409	829.9	1959.8	2789.7	2.283	4.186	6.469
200	15.5506	0.001157	0.1272	852.4	1940.4	2792.8	2.331	4.101	6.431
205	17.245	0.001164	0.1151	875.0	1920.6	2795.5	2.378	4.017	6.394
210	19.080	0.001173	0.1043	897.6	1900.2	2797.9	2.425	3.933	6.358
215	21.063	0.001181	0.09465	920.6	1879.3	2799.4	2.471	3.850	6.321
220	23.201	0.001190	0.08606	943.7	1857.8	2801.5	2.518	3.767	6.285
225	25.504	0.001199	0.07837	966.9	1835.6	2802.5	2.564	3.685	6.249
230	27.979	0.001209	0.07147	990.3	1812.9	2803.4	2.610	3.603	6.213
235	30.635	0.001219	0.06527	1013.8	1789.5	2803.4	2.656	3.522	6.178
240	33.480	0.001229	0.05967	1037.6	1765.5	2803.1	2.702	3.440	6.142
245	36.524	0.001240	0.05462	1061.6	1740.7	2802.2	2.748	3.359	6.107

T	p	v_f	v_g	h_f	h_{fg}	h_g	s_f	s_{fg}	s_g
250	39.776	0.001251	0.05006	1085.8	1715.1	2800.9	2.793	3.278	6.072
255	43.245	0.001263	0.04591	1110.2	1688.7	2799.0	2.839	3.197	6.037
260	46.940	0.001276	0.04215	1134.9	1661.5	2796.4	2.885	3.116	6.001
265	50.872	0.001289	0.03872	1159.9	1633.3	2793.3	2.931	3.035	5.966
270	55.051	0.001302	0.03560	1185.2	1604.2	2789.4	2.976	2.954	5.930
275	59.487	0.001317	0.03274	1210.8	1574.0	2784.9	3.022	2.872	5.894
280	64.192	0.001332	0.03013	1236.8	1542.8	2779.6	3.068	2.789	5.857
285	69.175	0.001348	0.02774	1263.1	1510.3	2773.4	3.114	2.706	5.820
290	74.449	0.001366	0.02554	1289.9	1476.4	2766.3	3.161	2.622	5.783
295	80.025	0.001384	0.02351	1317.1	1441.2	2758.3	3.208	2.537	5.744
300	85.92	0.001404	0.02164	1344.9	1404.3	2749.2	3.255	2.450	5.705
305	92.14	0.001425	0.01992	1373.2	1365.8	2738.9	3.302	2.362	5.665
310	98.70	0.001448	0.01832	1402.1	1325.2	2727.2	3.351	2.272	5.623
315	105.61	0.001472	0.01683	1431.7	1282.4	2714.1	3.400	2.180	5.580
320	112.90	0.001499	0.01545	1462.2	1237.5	2699.6	3.449	2.086	5.535
325	120.57	0.001529	0.01417	1493.5	1190.0	2683.5	3.500	1.989	5.489
330	128.65	0.001562	0.01297	1526.0	1139.5	2665.5	3.552	1.889	5.441
335	137.14	0.001599	0.01184	1559.7	1085.5	2645.2	3.605	1.785	5.390
340	146.08	0.001639	0.01078	1594.8	1027.2	2622.0	3.661	1.675	5.336
345	155.48	0.001686	0.00977	1631.8	963.6	2595.4	3.718	1.559	5.277
350	165.37	0.001741	0.00881	1671.2	893.0	2564.2	3.779	1.433	5.212
355	175.77	0.001807	0.00787	1713.9	813.6	2527.5	3.844	1.294	5.138
360	186.74	0.001894	0.00694	1761.5	719.6	2481.1	3.916	1.136	5.053
365	198.30	0.002016	0.00600	1817.5	603.4	2420.9	4.001	0.945	4.946
370	210.52	0.002225	0.00493	1892.4	438.4	2330.7	4.114	0.682	4.795
374	220.86	0.00280	0.00347	2031.8	114.6	2146.4	4.326	0.177	4.503
374.15	221.2	0.00317	0.00317	2084	0	2084	4.430	0	4.430

PROPERTIES OF SATURATED WATER AND STEAM : PRESSURE TABLE

p	T	v_f	v_g	h_f	h_{fg}	h_g	s_f	s_{fg}	s_g
0.010	6.98	0.001000	129.2	29.34	2484.3	2513.6	0.106	8.868	8.974
0.020	17.51	0.001001	67.01	73.45	2459.5	2533.0	0.261	8.462	8.723
0.030	24.10	0.001003	45.67	101.0	2444.0	2545.0	0.354	8.222	8.576
0.040	28.98	0.001004	34.81	121.4	2432.4	2553.9	0.422	8.051	8.473
0.050	32.90	0.001005	28.20	137.8	2423.1	2560.8	0.476	7.918	8.394
0.060	36.18	0.001007	23.74	151.5	2415.3	2566.8	0.521	7.808	8.329
0.070	39.02	0.001008	20.53	163.4	2408.5	2571.9	0.559	7.715	8.274
0.080	41.53	0.001009	18.10	173.9	2402.5	2576.4	0.593	7.635	8.227
0.090	43.79	0.001009	16.20	183.3	2397.1	2580.4	0.622	7.563	8.186
0.10	45.83	0.001010	14.67	191.8	2392.2	2584.1	0.649	7.499	8.149
0.15	54.00	0.001014	10.02	226.0	2372.5	2598.5	0.755	7.252	8.007
0.20	60.09	0.001017	7.648	251.5	2357.7	2609.1	0.832	7.075	7.907
0.25	64.99	0.001020	6.203	272.0	2345.7	2617.6	0.893	6.937	7.830
0.30	69.13	0.001022	5.228	289.3	2335.4	2624.8	0.944	6.823	7.767
0.35	72.71	0.001024	4.525	304.3	2326.5	2630.9	0.988	6.727	7.715
0.40	75.89	0.001026	3.992	317.7	2318.6	2636.3	1.026	6.643	7.669
0.45	78.74	0.001028	3.575	329.6	2311.4	2641.1	1.060	6.569	7.629
0.50	81.35	0.001030	3.239	340.6	2304.9	2645.4	1.091	6.502	7.593
0.60	85.95	0.001033	2.731	359.9	2293.2	2653.1	1.145	6.386	7.531
0.70	89.96	0.001036	2.364	376.8	2282.9	2659.7	1.192	6.287	7.478
0.80	93.51	0.001039	2.087	391.7	2273.7	2665.4	1.233	6.201	7.434
0.90	96.71	0.001041	1.869	405.2	2265.4	2670.6	1.270	6.125	7.394
1.00	99.63	0.001043	1.694	417.5	2257.7	2675.2	1.303	6.056	7.359
1.20	104.81	0.001047	1.428	439.4	2244.0	2683.3	1.361	5.937	7.298
1.40	109.32	0.001051	1.237	458.4	2231.9	2690.3	1.411	5.835	7.246
1.60	113.32	0.001054	1.091	475.4	2221.0	2696.4	1.455	5.747	7.202
1.80	116.93	0.001058	0.977	490.7	2211.1	2701.8	1.494	5.668	7.163

47

p	T	v_f	v_g	h_f	h_{fg}	h_g	s_f	s_{fg}	s_g
2.00	120.23	0.001061	0.8856	504.7	2201.9	2706.6	1.530	5.597	7.127
2.50	127.43	0.001067	0.7186	535.4	2181.6	2716.9	1.607	5.446	7.053
3.00	133.54	0.001073	0.6057	561.4	2163.9	2725.4	1.672	5.321	6.992
3.50	138.88	0.001079	0.5241	584.3	2148.2	2732.5	1.727	5.214	6.941
4.00	143.63	0.001084	0.4624	604.7	2133.9	2738.6	1.776	5.120	6.897
4.50	147.92	0.001088	0.4139	623.2	2120.8	2744.0	1.820	5.037	6.857
5.00	151.85	0.001093	0.3748	640.1	2108.6	2748.7	1.860	4.962	6.822
6.00	158.84	0.001101	0.3156	670.4	2086.4	2756.8	1.931	4.830	6.761
7.00	164.96	0.001108	0.2728	697.1	2066.4	2763.5	1.992	4.717	6.709
8.00	170.41	0.001115	0.2403	720.9	2048.2	2769.1	2.046	4.617	6.663
9.00	175.36	0.001121	0.2149	742.6	2031.2	2773.8	2.094	4.529	6.623
10.0	179.88	0.001127	0.1944	762.6	2015.3	2777.9	2.138	4.449	6.586
11.0	184.06	0.001133	0.1774	781.1	2000.4	2781.5	2.179	4.375	6.554
12.0	187.96	0.001139	0.1633	798.4	1986.2	2784.6	2.216	4.307	6.523
13.0	191.60	0.001144	0.1512	814.7	1972.6	2787.3	2.251	4.244	6.495
14.0	195.04	0.001149	0.1408	830.0	1959.6	2789.7	2.284	4.186	6.469
15.0	198.28	0.001154	0.1317	844.6	1947.1	2791.8	2.314	4.130	6.445
16.0	201.37	0.001159	0.1237	858.5	1935.6	2793.6	2.344	4.078	6.422
17.0	204.30	0.001163	0.1167	871.8	1923.4	2795.2	2.371	4.028	6.400
18.0	207.10	0.001168	0.1104	884.5	1912.1	2796.6	2.398	3.981	6.379
19.0	209.79	0.001172	0.1047	896.8	1901.1	2797.8	2.423	3.936	6.359
20.0	212.37	0.001177	0.09957	908.6	1890.4	2798.9	2.447	3.893	6.340
25.0	223.94	0.001197	0.07994	961.9	1840.4	2803.3	2.554	3.702	6.257
30.0	233.84	0.001216	0.06665	1008.3	1795.0	2803.3	2.645	3.541	6.186
35.0	242.54	0.001235	0.05705	1049.8	1753.0	2802.7	2.725	3.399	6.125
40.0	250.33	0.001252	0.04977	1087.4	1713.4	2800.8	2.797	3.273	6.070
45.0	257.41	0.001269	0.04405	1122.1	1675.7	2797.8	2.861	3.158	6.020

p	T	v_f	v_g	h_f	h_{fg}	h_g	s_f	s_{fg}	s_g
50.0	263.92	0.001286	0.03944	1154.5	1639.5	2794.0	2.921	3.053	5.973
55.0	269.94	0.001302	0.03563	1184.9	1604.6	2789.5	2.976	2.955	5.930
60.0	275.56	0.001319	0.03244	1213.7	1570.6	2784.3	3.027	2.862	5.890
65.0	280.83	0.001335	0.02972	1241.1	1537.5	2778.5	3.076	2.775	5.851
70.0	285.80	0.001351	0.02737	1267.4	1504.9	2772.3	3.122	2.692	5.814
75.0	290.51	0.001367	0.02532	1292.6	1472.9	2765.6	3.166	2.613	5.779
80.0	294.98	0.001384	0.02352	1317.0	1441.3	2758.3	3.207	2.537	5.744
85.0	299.24	0.001401	0.02192	1340.6	1410.1	2750.7	3.248	2.463	5.711
90.0	303.31	0.001417	0.02048	1363.5	1379.0	2742.5	3.286	2.392	5.679
95.0	307.22	0.001435	0.01919	1385.9	1348.0	2733.9	3.324	2.323	5.647
100	310.96	0.001452	0.01802	1407.7	1317.1	2724.8	3.360	2.255	5.615
110	318.04	0.001488	0.01598	1450.1	1255.4	2705.5	3.430	2.123	5.553
120	324.64	0.001527	0.01426	1491.4	1193.5	2684.7	3.496	1.997	5.493
130	330.81	0.001568	0.01278	1531.4	1131.0	2662.3	3.561	1.873	5.433
140	336.63	0.001612	0.01149	1571.0	1067.0	2638.0	3.623	1.750	5.373
150	342.12	0.001658	0.01035	1610.2	1001.1	2611.3	3.685	1.627	5.312
160	347.32	0.001710	0.00932	1649.7	931.9	2581.6	3.746	1.502	5.248
170	352.26	0.001769	0.00838	1690.0	858.4	2548.3	3.808	1.372	5.181
180	356.96	0.001838	0.00751	1731.8	778.6	2510.4	3.872	1.236	5.108
190	361.44	0.001923	0.00668	1776.5	689.2	2465.7	3.941	1.086	5.027
200	365.71	0.002039	0.00585	1826.6	583.9	2410.6	4.014	0.914	4.928
210	369.79	0.002213	0.00498	1888.5	447.1	2335.6	4.108	0.695	4.803
220	373.7	0.00269	0.00368	2007.9	170.1	2178.0	4.289	0.263	4.552
221.2	374.15	0.00317	0.00317	2084	0	2084	4.430	0	4.430

PROPERTIES OF SUPERHEATED AND SUPERCRITICAL STEAM

p	T	50	100	150	200	250	300	350	400	500	600	700	800
0.01	v	149.1	172.2	195.3	218.4	241.4	264.5	287.6	310.7	356.8	403.0	449.1	495.3
	h	2595	2689	2784	2880	2978	3077	3178	3280	3489	3706	3929	4159
	s	9.241	9.512	9.751	9.966	10.163	10.344	10.512	10.670	10.960	11.223	11.465	11.690
0.05	v	29.78	34.42	39.04	43.66	48.28	52.90	57.51	62.13	71.36	80.59	89.82	99.06
	h	2594	2688	2784	2880	2978	3077	3177	3280	3489	3706	3929	4159
	s	8.496	8.768	9.008	9.223	9.420	9.601	9.769	9.927	10.217	10.480	10.722	10.947
0.1	v	14.87	17.20	19.51	21.83	24.14	26.45	28.76	31.06	35.68	40.30	44.91	49.53
	h	2592	2688	2783	2880	2977	3077	3177	3280	3489	3706	3929	4159
	s	8.173	8.447	8.688	8.903	9.100	9.281	9.449	9.607	9.897	10.160	10.402	10.628
0.5	v		3.420	3.890	4.356	4.821	5.284	5.747	6.209	7.134	8.058	8.981	9.905
	h		2683	2780	2878	2976	3076	3177	3279	3489	3705	3929	4159
	s		7.694	7.940	8.158	8.355	8.537	8.705	8.864	9.154	9.417	9.659	9.885
1.0	v		1.696	1.937	2.173	2.406	2.639	2.871	3.103	3.565	4.028	4.490	4.952
	h		2676	2777	2876	2975	3075	3176	3278	3488	3705	3928	4159
	s		7.360	7.614	7.834	8.033	8.215	8.384	8.543	8.834	9.097	9.339	9.565
2.0	v			0.9602	1.081	1.199	1.316	1.433	1.549	1.781	2.013	2.244	2.475
	h			2770	2871	2971	3072	3174	3277	3487	3704	3928	4158
	s			7.280	7.507	7.708	7.892	8.062	8.221	8.513	8.776	9.019	9.244
3.0	v			0.6342	0.7166	0.7965	0.8754	0.9536	1.031	1.187	1.341	1.496	1.650
	h			2762	2866	2968	3070	3172	3275	3486	3703	3927	4158
	s			7.078	7.312	7.517	7.702	7.873	8.032	8.324	8.589	8.831	9.057
4.0	v			0.4710	0.5345	0.5953	0.6549	0.7139	0.7725	0.8893	1.005	1.121	1.237
	h			2753	2862	2965	3067	3170	3274	3485	3703	3927	4157
	s			6.929	7.172	7.379	7.566	7.738	7.898	8.191	8.455	8.698	8.924
5.0	v				0.4252	0.4745	0.5226	0.5701	0.6172	0.7108	0.8040	0.8969	0.9896
	h				2857	2962	3065	3168	3272	3484	3702	3926	4157
	s				7.060	7.271	7.460	7.633	7.793	8.087	8.351	8.595	8.820

p	T	50	100	150	200	250	300	350	400	500	600	700	800
6.0	v				0.3522	0.3940	0.4344	0.4743	0.5136	0.5919	0.6697	0.7471	0.8245
	h				2851	2958	3062	3166	3270	3483	3701	3925	4157
	s				6.968	7.182	7.373	7.546	7.707	8.001	8.267	8.510	8.736
7.0	v				0.3001	0.3364	0.3714	0.4058	0.4397	0.5069	0.5737	0.6402	0.7065
	h				2846	2955	3060	3164	3269	3482	3700	3925	4156
	s				6.888	7.106	7.298	7.473	7.634	7.929	8.195	8.438	8.665
8.0	v				0.2610	0.2933	0.3242	0.3544	0.3842	0.4432	0.5018	0.5600	0.6181
	h				2840	2951	3057	3162	3267	3481	3699	3924	4156
	s				6.817	7.040	7.233	7.409	7.571	7.866	8.132	8.376	8.603
9.0	v				0.2305	0.2597	0.2874	0.3144	0.3410	0.3937	0.4458	0.4976	0.5493
	h				2835	2948	3055	3160	3266	3480	3699	3924	4155
	s				6.753	6.980	7.176	7.352	7.515	7.811	8.077	8.321	8.548
10.0	v				0.2061	0.2328	0.2580	0.2825	0.3065	0.3540	0.4010	0.4477	0.4943
	h				2829	2944	3052	3158	3264	3478	3698	3923	4155
	s				6.695	6.926	7.124	7.301	7.464	7.761	8.028	8.272	8.499
15.0	v				0.1324	0.1520	0.1697	0.1865	0.2029	0.2351	0.2667	0.2980	0.3292
	h				2796	2925	3039	3148	3256	3473	3694	3920	4152
	s				6.452	6.711	6.919	7.102	7.268	7.569	7.838	8.083	8.310
20.0	v					0.1115	0.1255	0.1386	0.1511	0.1756	0.1995	0.2232	0.2466
	h					2904	3025	3138	3248	3467	3690	3917	4150
	s					6.547	6.768	6.957	7.126	7.431	7.701	7.948	8.176
25.0	v					0.08704	0.09891	0.1097	0.1200	0.1399	0.1592	0.1783	0.1971
	h					2882	3011	3127	3239	3462	3686	3914	4148
	s					6.411	6.646	6.841	7.014	7.322	7.595	7.842	8.071
30.0	v					0.07061	0.08115	0.09051	0.09928	0.1161	0.1324	0.1483	0.1641
	h					2858	2995	3117	3231	3456	3682	3911	4146
	s					6.289	6.541	6.744	6.921	7.233	7.507	7.756	7.985

T p		50	100	150	200	250	300	350	400	500	600	700	800
35.0	v					0.05873	0.06842	0.07675	0.08447	0.09910	0.1132	0.1269	0.1405
	h					2831	2980	3105	3222	3450	3678	3908	4143
	s					6.176	6.449	6.659	6.840	7.156	7.433	7.682	7.912
40.0	v						0.05884	0.06642	0.07334	0.08635	0.09877	0.1109	0.1228
	h						2963	3094	3214	3445	3674	3905	4141
	s						6.364	6.584	6.769	7.089	7.368	7.618	7.849
45.0	v						0.05135	0.05837	0.06469	0.07643	0.08758	0.09842	0.1091
	h						2946	3082	3205	3439	3670	3902	4139
	s						6.286	6.515	6.705	7.029	7.310	7.561	7.793
50.0	v						0.04532	0.05192	0.05776	0.06850	0.07862	0.08844	0.09807
	h						2927	3070	3196	3433	3666	3899	4136
	s						6.212	6.451	6.646	6.975	7.258	7.510	7.743
60.0	v						0.03616	0.04220	0.04734	0.05659	0.06519	0.07347	0.08157
	h						2887	3045	3177	3421	3657	3893	4132
	s						6.071	6.336	6.541	6.879	7.166	7.421	7.655
70.0	v						0.02945	0.03522	0.03989	0.04808	0.05559	0.06278	0.06978
	h						2841	3018	3158	3410	3649	3887	4127
	s						5.934	6.231	6.448	6.796	7.088	7.345	7.581
80.0	v						0.02423	0.02994	0.03428	0.04170	0.04839	0.05476	0.06094
	h						2787	2990	3139	3398	3641	3881	4122
	s						5.793	6.133	6.364	6.723	7.019	7.279	7.516
90.0	v							0.02578	0.02991	0.03673	0.04279	0.04852	0.05406
	h							2959	3118	3385	3633	3874	4118
	s							6.039	6.286	6.657	6.958	7.220	7.458
100	v							0.02241	0.02639	0.03275	0.03831	0.04353	0.04856
	h							2926	3097	3373	3624	3868	4113
	s							5.947	6.213	6.596	6.902	7.166	7.406

p	T	50	100	150	200	250	300	350	400	500	600	700	800
125	v							0.01611	0.02000	0.02558	0.03025	0.03456	0.03866
	h							2828	3040	3341	3603	3853	4102
	s							5.714	6.043	6.461	6.780	7.051	7.294
150	v							0.01146	0.01566	0.02078	0.02487	0.02857	0.03207
	h							2693	2977	3309	3581	3837	4090
	s							5.443	5.883	6.345	6.677	6.954	7.201
175	v								0.01246	0.01735	0.02103	0.02430	0.02736
	h								2904	3275	3559	3821	4078
	s								5.724	6.239	6.586	6.870	7.122
200	v								0.00995	0.01477	0.01815	0.02110	0.02383
	h								2819	3239	3537	3806	4067
	s								5.556	6.142	6.505	6.796	7.051
300	v								0.002818	0.008684	0.01143	0.01364	0.01561
	h								2157	3084	3445	3742	4020
	s								4.482	5.795	6.234	6.557	6.829
400	v								0.001914	0.005618	0.008087	0.009932	0.01152
	h								1935	2906	3348	3677	3974
	s								4.119	5.474	6.014	6.371	6.662
500	v								0.001733	0.003882	0.006108	0.007725	0.009079
	h								1879	2722	3249	3612	3928
	s								4.009	5.176	5.821	6.214	6.524
600	v								0.001636	0.002950	0.004830	0.006673	0.007466
	h								1848	2571	3152	3548	3883
	s								3.939	4.937	5.648	6.077	6.405
700	v								0.001570	0.002466	0.003969	0.005257	0.006326
	h								1829	2468	3062	3486	3839
	s								3.886	4.769	5.494	5.955	6.300

P	T		50	100	150	200	250	300	350	400	500	600	700	800
800	v									0.001520	0.002191	0.003378	0.004518	0.005484
	h									1815	2400	2983	3428	3797
	s									3.842	4.651	5.361	5.845	6.206
900	v									0.001481	0.002016	0.002964	0.003964	0.004841
	h									1805	2353	2916	3373	3756
	s									3.805	4.563	5.248	5.746	6.120
1000	v									0.001448	0.001894	0.002668	0.003542	0.004338
	h									1798	2319	2860	3324	3718
	s									3.773	4.493	5.153	5.656	6.042

8.9   The standard atmosphere*

Symbols and Units

quantity	symbol	unit
height	H	m
temperature	T	$^{\circ}C$
velocity of sound	a	m/s
pressure	p	mb
pressure at ground level	p_o	mb
density	ρ	kg/m^3
density at ground level	ρ_o	kg/m^3
viscosity	μ	kg/ms

The table is based on the ISO Standard Atmosphere (1973).

---

* Reproduced by courtesy of "U.S. Standard Atmosphere", 1976.
Published by: National Oceanic and Atmospheric Administration
and National Aeronautics and Space Administration and The U.S.
Airforce. U.S. Government Printing Office, Washington,
DC 20402.

H	T	a	p	p/p_o	ρ	ρ/ρ_o	$(\rho/\rho_o)^{\frac{1}{2}}$	$\mu \times 10^5$
0	15.00	340.3	1013.25	1.0000	1.2250	1.0000	1.0000	1.7894
500	11.75	338.4	954.61	0.9421	1.1673	0.9529	0.9762	1.7737
1000	8.50	336.4	898.75	0.8870	1.1116	0.9075	0.9526	1.7578
1500	5.25	334.5	845.56	0.8345	1.0581	0.8637	0.9294	1.7419
2000	2.00	332.5	794.95	0.7846	1.0065	0.8216	0.9064	1.7260
2500	-1.25	330.6	746.83	0.7371	0.9569	0.7811	0.8838	1.7099
3000	-4.50	328.6	701.09	0.6919	0.9091	0.7421	0.8615	1.6937
3500	-7.75	326.6	657.64	0.6490	0.8632	0.7047	0.8311	1.6775
4000	-11.00	324.6	616.40	0.6083	0.8191	0.6687	0.8177	1.6611
4500	-14.25	322.6	577.28	0.5697	0.7768	0.6341	0.7963	1.6447
5000	-17.50	320.5	540.20	0.5331	0.7361	0.6009	0.7752	1.6281
5500	-20.75	318.5	505.07	0.4985	0.6971	0.5691	0.7544	1.6115
6000	-24.00	316.5	471.81	0.4656	0.6597	0.5385	0.7338	1.5947
6500	-27.25	314.4	440.35	0.4346	0.6238	0.5093	0.7137	1.5779
7000	-30.50	312.3	410.61	0.4052	0.5895	0.4812	0.6937	1.5610
7500	-33.75	310.2	382.51	0.3775	0.5566	0.4544	0.6741	1.5439
8000	-37.00	308.1	356.00	0.3513	0.5252	0.4287	0.6548	1.5268

H	T	a	p	p/p_0	ρ	ρ/ρ_0	$(\rho/\rho_0)^{\frac{1}{2}}$	$\mu \times 10^5$
8500	-40.25	305.9	330.99	0.3267	0.4951	0.4042	0.6358	1.5095
9000	-43.50	303.8	307.42	0.3034	0.4664	0.3807	0.6170	1.4922
9500	-46.75	301.6	285.24	0.2815	0.4389	0.3583	0.5986	1.4747
10000	-50.00	299.5	264.36	0.2609	0.4127	0.3369	0.5804	1.4571
10500	-53.25	297.3	244.73	0.2415	0.3877	0.3165	0.5626	1.4394
11000	-56.50	295.1	226.32	0.2234	0.3639	0.2971	0.5451	1.4216
11500	-56.50	295.1	209.16	0.2064	0.3363	0.2746	0.5240	1.4216
12000	-56.50	295.1	193.30	0.1908	0.3108	0.2537	0.5037	1.4216
12500	-56.50	295.1	178.65	0.1763	0.2873	0.2345	0.4843	1.4216
13000	-56.50	295.1	165.10	0.1629	0.2655	0.2167	0.4655	1.4216
13500	-56.50	295.1	152.59	0.1506	0.2454	0.2003	0.4475	1.4216
14000	-56.50	295.1	141.02	0.1392	0.2268	0.1851	0.4302	1.4216
14500	-56.50	295.1	130.33	0.1286	0.2096	0.1711	0.4136	1.4216
15000	-56.50	295.1	120.45	0.1189	0.1937	0.1581	0.3976	1.4216
15500	-56.50	295.1	111.31	0.1099	0.1790	0.1461	0.3822	1.4216
16000	-56.50	295.1	102.87	0.1015	0.1654	0.1350	0.3674	1.4216
17000	-56.50	295.1	87.87	0.0867	0.1413	0.1153	0.3396	1.4216
18000	-56.50	295.1	75.05	0.0741	0.1207	0.0985	0.3138	1.4216
19000	-56.50	295.1	64.10	0.0633	0.1031	0.0841	0.2900	1.4216
20000	-56.50	295.1	54.75	0.0540	0.0880	0.0719	0.2681	1.4216

9. MECHANICS AND STRUCTURES

9.1 Material constants

quantity	symbol
density	ρ
Young's modulus	E
modulus of rigidity	G
bulk modulus	K
Poisson's ratio	ν
coefficient of linear expansion	α
coefficient of volumetric expansion	β

Some typical values are as follows:

		mild steel	aluminium alloy	brass	concrete	water
ρ	(kg/m^3)	7850	2720	8410	2400	1000
	(lb/ft^3)	490	170	525	150	62.4
E	(kN/mm^2)	207	68.9	103	13.8	
	(lbf/in^2)	30×10^6	10×10^6	15×10^6	2×10^6	
G	(kN/mm^2)	79.6	26.5	38.3		
	(lbf/in^2)	11.5×10^6	3.85×10^6	5.56×10^6		
K	(kN/mm^2)	172	57.5	115		2.15
	(lbf/in^2)	25×10^6	8.33×10^6	16.7×10^6		3.1×10^5
ν		0.3	0.3	0.35	0.1	
α	(per $^\circ$C)	11×10^{-6}	23×10^{-6}	19×10^{-6}		
β	(per $^\circ$C)					2.1×10^{-4}

Relations between elastic constants

$$G = E/2(1+\nu) \qquad\qquad K = E/3(1-2\nu)$$

58

9.2 Beams bent about principal axis

	end slope	end deflection	central deflection

Cantilever with end moment M:

$$\frac{ML}{EI} \qquad \frac{ML^2}{2EI}$$

Cantilever with end point load W:

$$\frac{WL^2}{2EI} \qquad \frac{WL^3}{3EI}$$

Cantilever with uniformly distributed load W:

$$\frac{WL^2}{6EI} \qquad \frac{WL^3}{8EI}$$

Simply supported beam with end moments M:

$$\frac{ML}{2EI} \qquad\qquad\qquad \frac{ML^2}{8EI}$$

Simply supported beam with central point load W:

$$\frac{WL^2}{16EI} \qquad\qquad\qquad \frac{WL^3}{48EI}$$

Simply supported beam with uniformly distributed load W:

$$\frac{WL^2}{24EI} \qquad\qquad\qquad \frac{5WL^3}{384EI}$$

9.3   Stress and strain transformations

Two-dimensional stress system

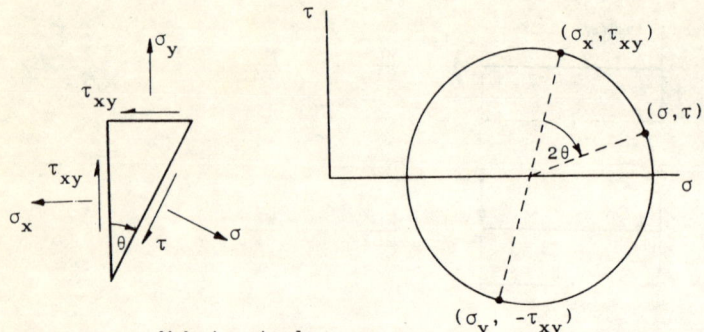

Mohr's circle

$$\sigma = \tfrac{1}{2}(\sigma_x + \sigma_y) + \tfrac{1}{2}(\sigma_x - \sigma_y)\cos 2\theta + \tau_{xy} \sin 2\theta$$

$$= \tfrac{1}{2}(\sigma_x + \sigma_y) + \tau^{max} \cos 2(\psi - \theta)$$

$$\tau = -\tfrac{1}{2}(\sigma_x - \sigma_y)\sin 2\theta + \tau_{xy} \cos 2\theta$$

$$= \tau^{max} \sin 2(\psi - \theta)$$

$$\text{where} \quad \tau^{max} = \left[\left(\frac{\sigma_x - \sigma_y}{2} \right)^2 + \tau_{xy}^2 \right]^{\frac{1}{2}}$$

$$\tan 2\psi = 2\tau_{xy}/(\sigma_x - \sigma_y)$$

Two-dimensional strain system

ε_x, ε_y, ε = direct strains 'corresponding' to σ_x, σ_y, σ

γ_{xy}, γ = shear strains 'corresponding' to τ_{xy}, τ

$$\varepsilon = \tfrac{1}{2}(\varepsilon_x + \varepsilon_y) + \tfrac{1}{2}(\varepsilon_x - \varepsilon_y)\cos 2\theta + \tfrac{1}{2}\gamma_{xy}\sin 2\theta$$

$$\tfrac{1}{2}\gamma = -\tfrac{1}{2}(\varepsilon_x - \varepsilon_y)\sin 2\theta + \tfrac{1}{2}\gamma_{xy}\cos 2\theta$$

Three-dimensional stress system

If the principal stresses are σ_1, σ_2, σ_3, the principal shear stresses are $(\sigma_1 - \sigma_2)/2$, $(\sigma_2 - \sigma_3)/2$ and $(\sigma_3 - \sigma_1)/2$.

Strain energy per unit volume U may be expressed as

$$U = (\sigma_1 + \sigma_2 + \sigma_3)^2/18K$$
$$+ [(\sigma_1 - \sigma_2)^2 + (\sigma_2 - \sigma_3)^2 + (\sigma_3 - \sigma_1)^2]/12G$$